WITHDRAWN

SOCIAL AND NATURAL SCIENCES: THE ISLAMIC PERSPECTIVE

ISLAMIC EDUCATION SERIES
Other Titles in the Series

Crisis in Muslim Education
Aims and Objectives of Islamic Education
Philosophy Literature and Fine Arts
Curriculum and Teacher Education
Education and Society in the Muslim World
Muslim Education in the Modern World

General Editor   Syed Ali Ashraf

SOCIAL AND NATURAL SCIENCES: THE ISLAMIC PERSPECTIVE

edited by

Professor Isma'il R. Al-Faruqi
and
Dr Abdullah Omar Nasseef

HODDER AND STOUGHTON
KING ABDULAZIZ UNIVERSITY, JEDDAH

British Library Cataloguing in Publication Data

Social and natural sciences. – (Islamic education series).
  1. Islam – Education
  2. Social sciences – Study and teaching – Islamic countries
  3. Science – Study and teaching – Islamic countries
  I. Nasseef, Abdullah Omar   II. al Fārūqī Ismáil Rāgī   III. Series
  300'.7'1017671   LC903

ISBN 0 340 23613 2

First printed 1981
Copyright © 1981 King Abdulaziz University, Jeddah

All rights reserved. No part of this publication may be reproduced or transmitted in any form or by any means, electronic or mechanical, including photocopy, recording, or any information storage or retrieval system, without permission in writing from the publisher.

Filmset, printed and bound in Great Britain for Hodder and Stoughton Educational, a division of Hodder and Stoughton Ltd., Mill Road, Dunton Green, Sevenoaks, Kent by Hazell Watson & Viney Ltd, Aylesbury, Bucks.

Foreword

It is becoming increasingly apparent that the Muslim world occupies a unique position between two conflicting ideological groups: the liberal West and the Communist World. Muslims can play a vital role in maintaining peace in this world and in giving a lead to all humanity if they demonstrate unity, discipline and wisdom. Unity can be achieved only through their complete submission to Faith. Intellectual, moral and physical discipline can be attained only if they act according to their beliefs. Wisdom can be acquired only when their intellect, soul and body obey the commands of Spirit and move freely within the orbits laid down by Allah through Prophet Muhammad, peace and blessings of Allah be on him. Education is the only means through which such wisdom can be acquired. Unfortunately the confusion of ideologies in the West and the materialistic dogma of the Communist World are invading the Muslim World. In order to achieve quick intellectual and material progress the Muslim World has accepted the Western system of education and tried its best to blend the Islamic system with that of the West. But the compromise has proved to be odious because Faith and secularism cannot be synthesized. Muslim scholars must therefore evolve Islamic schools of thought for all branches of knowledge: the want of such schools is causing the Muslim world to divide into conflicting groups even within each individual country.

The problem is most acute in the sphere of social sciences. As these sciences influence not only individuals in their personal private thought and action but the whole society, they create an environment and complex economic, political and social infrastructures which are difficult to alter, modify or remove. It is high time therefore that we substituted the secularist metaphysics by Islamic concepts and built up an Islamic infrastructure to replace the already entrenched secularist system.

In natural sciences the philosophical concepts and the methodology which deny the presence of Allah's will in this creation, are creating an anti-Islamic attitude. A new metaphysics and a modified methodology with a drastically Faith-oriented approach are long overdue.

I regard the publication of this volume as a significant event in Muslim scholarship. It consists of key papers submitted at the First World Conference on Muslim education. I am sure this book will elicit a creative response from scholars throughout the world.

Muhammad Omar Zubair
President, First World Conference on Muslim Education
Ex-President, King Abdulaziz University

Contents

Foreword		v
Part 1	SOCIAL SCIENCES	
Preface		3
Introduction		5
Chapter One	Islamizing the Social Sciences — *Isma'il R. Al-Faruqi*	8
Chapter Two	Sociology and Muslim Social Realities — *Ilyas Ba-Yunus*	21
Chapter Three	An Islamic Concept of History — *Abdul Hamid Siddiqui*	41
Chapter Four	On Islamizing the Discipline of Psychology — *Abdul Hamid al-Hashimi*	49
Chapter Five	Restructuring the Study of Economics in Muslim Universities — *Mohammad Nejatullah Siddiqui*	71
Chapter Six	Beyond the Muslim Nation-States — *Kalim Siddiqui*	87
Chapter Seven	The Ummah and its Civilizational Crisis — *Abdul Hamid Abu Suleiman* — and translated into English from the Arabic by *Isma'il R. Al-Faruqi*	100
Chapter Eight	Elements for an Islamic Anthropology — *Saibo Mohamed Mauroof*	116

Part 2   NATURAL SCIENCES

Preface   143
Introduction   145
Chapter One   Humanistic Social Sciences —Studies
               in Higher Technical Education.
               —*S. Waqar Ahmed Husaini*   148
Chapter Two   Scientific Education in Muslim
               Countries —Principles and Guidelines
               —*Ata-ur-Rahman*   167

Part 1 Social Sciences
Edited by Professor Isma'il R. Al-Faruqi

Preface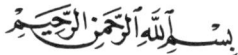

Formulating Islamic concepts in social sciences as substitutes for those of the Western world is not an easy task. Not only are modern educated Muslims accustomed to Western concepts, Western thought, Western methods of approach to these subjects, as well as methods of teaching them, they are conditioned by their upbringing and the secular modern education that they have received. A reassessment of all branches of social sciences is therefore a necessity. But any revision along Islamic lines is possible only when faith in the validity of Quranic statements and the Prophet's sayings is firmly ingrained in the souls of the planners and they accept without qualification these Quranic premises. If the Quranic concept of the relationship between morality and historical events and the Quranic narration of the past are ignored, a Muslim historian cannot attempt to formulate an Islamic concept of history. Similarly, an Islamic school of sociology cannot emerge if Muslim sociologists do not analyse their data with reference to the Quranic concept of absolute values and human rights, duties and obligations. In the same way, it is impossible to create an Islamic school of psychology if the Quranic concept of human nature, including the idea of the relationship between body, self and spirit, and man's destiny as the vicegerent of God on earth, is overlooked, evaded or ignored. The same is true for economics and political science because neither the capitalist, interest-controlled, economic system, nor the rigid regimentation of the Marxist economy and political system, can be acceptable to a Muslim.

Muslim scholars must therefore draw their basic principles from the Quran, and the Sunnah, and with reference to those principles interpret man (and his activities that fall within social sciences) so laying the foundation of an Islamic school of social sciences. In the social sciences section of this book an attempt has been made to initiate research in this field. But to make future attempts successful, the fundamental requirement is a metaphysics to integrate faith and rational thinking. At present there exists a complete divorce between the two: a divorce that has allowed secularism to flourish. It is the concept of human nature from the Islamic point of view which must be re-established, a viewpoint which is simultaneously both natural and spiritual. Only then can a goal for humanity be accepted and asserted, and all other conceptual formulations become possible. We hope that a new generation of Muslim scholars will emerge to carry on this task and guide this present attempt to successful fruition.

Syed Ali Ashraf
General Editor

Introduction

Islam is a vision of world, time and life which God has revealed to humankind by a succession of messengers, beginning with Adam (*Alayhi al salam*) and ending with Muhammad (*Salla Allahu alayhi wa sallam*). Repeatedly, the vision has been clouded through transmission from person to person, generation to generation and people to people, because of human error, prejudice and passion, and the ever-changing processes which constitute human history. Every new revelation reaffirmed the vision, clarified it and re-crystallized it in terms appropriate to the new conditions of history. Every revelation contained a "what," the goals, objectives and values of human life and existence, and a "how", the concrete applications of the "what" to the needs of everyday life. Thus content and form of revelation were bound in indissoluble unity in the law codes of Lippit Ishtar, of Sargon of Akkad, of Hammurabi, of Ibrahim, Ismail, Ishaq and Yaqub, Ayyub, of Musa and Isa (*Alayhim al salam*), and all of the other prophets, "some of whom we have told you (O Muhammad); and others about whom we have not. (Quran 4:163; 40:78).

The precious gift of revelation promised to Adam and posterity (Quran 2:28) was an act of mercy by the All-Merciful, Ever-Beneficent God. Its purpose is to help man achieve *falah* or felicity. With further mercy, the Creator—*subhanahu wa ta'ala*—has equipped all humans with a *sensus communis* with which to perceive His divine presence; and a natural reason with which to understand the relevance, and hence normality, of God's will or commandments, to the affairs of life. That which is revealed by God to His prophets is discoverable by reason in the world of creation as well as internally "within your own souls" (Quran 51:21). Both reason and revelation are avenues of knowledge, each designed to correct not the truth (the truth stands beyond correction!) but our understanding of the truth. Where our rational knowledge is deficient or incoherent, revelation is the only recourse. Where our understanding of revelation is marred by forgetfulness, prejudice and passion, the only recourse is to reason. However, reason must be understood in its broadest sense of reasonableness. Hence, it is inclusive of theoretical as well as axiological reality. It is not to be taken in the narrow sense of the philosophical "rationalist", whether scholastic, deistic, idealistic, materialistic or positivistic.

Being the last and final, the Muhammadan revelation has separated the "what" from the "how", the vision and its reality. The Quran and the Sunnah embody the former. They have, equally, laid down detailed prescriptive legislation in the departments of rituals, marriage and divorce, inheritance and economic life, binding their forms forever. In other areas of human activity, they determined the values and general principles which should guide action, and left the figurization and concretization for humans to work out in the spirit of their own times, in relation to their historical settings. To add still more mercy, the Almighty has, in His wisdom, revealed to us in both sources the methodological principles for the renewal, the *Vergegenwartigung* of the Sharia or law of Islam, entrusting the task of renewing the Sharia, of making it presently-relevant to human conditions, and of rekindling the vision which the Prophet (peace and blessings of God be on him) imparted to his companions (may God be pleased with them), to man. This is why the Sharia is both divine and human: the former by virtue of it embodying the divine principles, whether of content or of method; the latter by virtue of its translation of those principles into the prescription for the age.

In the past, the *Ummah* or world community of Islam has trusted and relied upon the 'Ulama or "men of knowledge" to guide its course in history. These have been true to the trust. In the early formative period, they figurized the vision that the Prophet Muhammad (SAAS) had imparted, and translated it into prescriptive directions for everyday living. In the middle period, they guarded the vision, and its figurization. In modern times their task has become difficult and challenging. The *Ummah* has been torn between conservatives who seek to preserve intact the figurization of earlier times, and liberals anxious to alter the figurization as present realities dictate. Modern times have increased tremendously the complexity of human life, necessitating a great amount of preparation for a successful governance of life. As guides and leaders endowed with clear perception of the Muhammadan vision, of the wisdom the fathers displayed in their figurizations, and of the knowledge of modern realities, the Muslim social scientists are the 'Ulama of the *Ummah* today. They are the planners of its strategies and designers of its future, the educators of the *Ummah* at large as well as of its political, social and economic leadership. In short, they are the scientists whose object of study is the

Ummah in all its activities as an *Ummah*. Their studies are the "ummatic" sciences, i.e., those disciplines which study human behaviour as it affects, and is affected by, society. The significance of ummatic science is fulfillment of its ultimate responsibility for the *Ummah*'s course in history. As such, the Muslim social scientist is student and teacher at the same time. Both the Ummah's vision and its pursuit constitute his concern as *alim* (man of knowledge) and Muslim (committed to the vision). As the *Ummah* is the carrier of the divine message and witness unto the rest of humanity (Quran 2:143), so is the Muslim social scientist the trustee of the vision and its first executor. He is, in a unique sense, the true heir of the Prophet (SAAS) in the role of witness of God over the Ummah (Quran 2:142).

Such responsibility has determined the First International Conference of Islamic Education, held at Makkah al Mukarramah in 1397/1977. The present book is an anthology of some of the ideas presented to it at the meeting of its Social Sciences Committee.

Temple University Isma' il R. Al-Faruqi
Philadelphia, Pennsylvania

Chapter One
Islamizing the Social Sciences

Isma'il Ragi Al-Faruqi

Isma'il Ragi Al-Faruqi, born in Palestine, is Professor of Islamics at Temple University, Philadelphia. He studied at the American University of Beirut, Indiana, and Harvard and did postdoctoral work on Islam at al-Azhar University in Cairo and in Christianity and Judaism at McGill University in Montreal as a Rockefeller Foundation Fellow.

Professor Al-Faruqi has held teaching posts at the Institute of Islamic Studies at McGill University, Montreal; the Central Institute of Islamic Research, Karachi; the Institute of Higher Arabic Studies of the League of Arab States, Cairo; and Syracuse University. He has held visiting professorships at Cairo University, Azhar University, and Alexandria University.

His publications include a number of articles in *The Canadian Journal of Theology*, *The Journal of the American Academy of Religion*, *The Bulletin of the Faculty of Arts of Cairo University*, *The Muslim World*, *Numen*, *Zygon*, and other periodicals. Besides a number of English translations, of which the latest is Haykal's *The Life of Muhammad* (Philadelphia: Temple University Press, 1973), he is the author of *Urubah and Religion* (Amsterdam: Djambatam, 1961), *Christian Ethics* (Montreal: McGill University Press, 1966), *Particularism in the Old Testament and Contemporary Sects in Judaism* (Cairo: League of Arab States, 1963, 68), co-author of *The Great Asian Religions* (New York: Macmillan, 1969) and co-editor of *The Historical Atlas of Religions of the World* (New York: Macmillan, 1974).

I. The Rise of the Social Sciences in the West

The disciplines which the West calls "the social sciences" are barely a century old. In most universities, these include five disciplines, sociology, anthropology, political science, economics and history. Two more disciplines enjoy a double status. They are geography and psychology. If geography undertakes to help any of the other social sciences by undertaking to relate the findings of that discipline to space, it becomes another social science and is called social, human, political, economic, historical or cultural geography. Otherwise geography studies the earth, as the etymology of its name indicates, and is classified with the natural sciences. In this case it is called "physical" or "natural" geography. Likewise is the case of psychology. If it studies the individual, it is classified with the natural sciences; but if it studies the group, then it becomes a social science and is called "social" psychology.

Although these disciplines have achieved autonomous status in the universities only during the last century, the forces which led to their emergence and success are about two centuries old. The seventeenth and eighteenth century rationalists built a great system of thought by which they sought to reestablish the first premises of Western Christian culture on a rational basis. Most Western Europeans looked upon the French Revolution as the visible movement which concretized and sought to universalize that system. Its subsequent failure was both the cause and the effect of an ingrown scepticism which for centuries before waged a relentless battle against the magisterium of the Church. This prolonged and merciless war was fought in the fields of the natural sciences, primarily astronomy and cosmology, where the Church had held, along with scripture, opinions which ran counter to the observations of empirical science. This scepticism succeeded in knocking out the cosmological doctrines of the Church, as well as the rational systems which in spite of their rationalism, still held as true the major premises of Christianity.

It is not surprising that this easy victory gave the new tendency self-confidence and faith in its empiricism, in its persistent questioning, in its rejection of all deductive reasoning, and in its total reliance on the methodology of natural science. And it was to this that Auguste Comte could look back and could see that the world had been wallowing in the mud and naiveté of belief and had only recently awakened to the dictates of reason. In his own time, he saw the shortlived age of reason as one in which Europe succeeded in terminating the intellectual tyranny of the Church. But looking ahead, he optimistically foresaw a new age dawning, the age of "positive science", in which man, now finally liberated from naive faith as well as dogmatic reason, would assume his place in a world from which both had been banished. Man, henceforth, was according to Comte, to apply science in the governance of his own life. Just as he had applied the science of medicine to his individual physical existence, he should begin to apply social science to his social behaviour and to that of the group as a whole. According to the father of social science in the West, natural science possessed the unquestionable methodology of truth. Since it had been successfully applied in physical nature, it was now the time to extend this success to the realm of human relations.

II. *The Prototype of Natural Science*

The core of this whole development was in the inductive methodology of natural science. The data of natural science are observable by the senses, isolable from one another, and measurable by the senses. They are "dead" in the sense that they are immune to the disposition of the observer. They reflect the same features and behaviour at all times as long as their own conditions remain unchanged, and are regardless of any subjective determinant of the observer. In science, no principle is sacrosanct and everything is questionable. The evidence of the experiment alone serves as a base for the hypothesis which remains valid as long as no other experiment has disproved it. The hypothesis constitutes a law of nature when repeated experiment and observation have confirmed its validity.

This theory carried the further appeal of making the whole world subject to explanation. Nature is finally forced to unlock her doors as science prises them open with its investigation of the causes and effects of phenomena. To discover the sufficient cause of phenomena, that is, to identify and lay bare for sense observation and measurement the determinants capable of bringing it about, is to explain it. Indeed, it is to make possible its control and engineering. Natural science is the key to the mastery of the world. Its capacity to understand is a capacity at once to engineer and to control. Hence it gives power as well as happiness. This vision of science gripped Western man and released tremendous energies for the exploration and usufruct of nature. The gains immediately won and achieved confirmed the vision beyond doubt and made science the undisputed avenue to utopia.

Naturally, what was possible in nature was then assumed to be possible in humanity, in the individual as well as in society, for both are nature and must be subject, if not to the same laws, to the same method of discovering them. Behind this certitude stood the hope of manipulating social reality to preconceived ends. Society was not moving as fast as scientific development. If the optimism generated by science was to stand, society would have to make faster changes. The question, how to bring such changes about, pressed the European mind to seek to understand social behaviour in order to direct its movement. It was over-hastily thought that the analysis of natural science could bring about such understanding and lay society open to control, to planning and direction.

III. The Shortcomings of Western Methodology

A. *The Denial of Relevance to A Priori Data*

The Western student of human nature and society was not in the mood to realize that not all the pertinent data of human behaviour are observable by the senses and hence subject to quantification and measurement. The human phenomenon does not consist of "natural" elements exclusively. Elements of a different order, the order of morality and spirit, enter into and determine it to a pre-eminent degree. These are not necessary corollaries of the elements of nature and are not deducible from them. They are autonomous in the sense of being valid in themselves even if the accompanying elements of nature differ from and violate them. And yet, no description of social relations is complete without reference to them. Nor are they universally the same in human groups but depend upon traditions of culture, religion and personal and group preference, which can never be exhaustively defined. Being spiritual, these elements are not isolable, separable from their natural carriers. Nor are they ever subject to the only measurement science knows, the quantitative. Science treated them as inexistent or irrelevant. It pressed the claims of its analysis of the observable, natural elements in human behaviour and, insisting on explaining human conduct, brought us bungled theories and incomplete explanations. In order that the analysis might remain scientific, the social scientist illegitimately reduced the moral and/or spiritual component of social reality to its material effect or carrier. His methodology remains to this day devoid of tools by which to recognize and deal with the spiritual.

B. *False Sense of Objectivity*

This elemental mistake in defining and identifying the data of social science led to another, namely, that any observer could establish the laws governing social reality if and only if he followed meticulously the rules of science. He should be careful to silence all personal biasses, suspend all prejudgment and allow the facts to speak for themselves. It was thought that under such rigour, the facts could not but unlock their secrets and thus subject themselves to scientific manipulation.

However, unlike those of natural science, the data of human behaviour are not dead, but alive. They are not impervious to the attitudes and preferences of the observer. They do not reveal themselves as they really are to each and every investigator. Attitudes, feelings, desires, judgments and hopes of men and women tend to shut themselves off to the observer devoid of sympathy for them. The explanation of this seeming discrimination by the data against the observer is to be found in the analysis of axiological perception. In the perception of "dead" objects, the senses of the observer are passive; they are totally determined by the data. In the perception of values, *per contra*, the observer actively empathizes or "emotes" with the data, whether for them or against them. Value-perception is itself value-determination, i.e., it takes place only when value is apprehended in actual experience. In other words, a value is said to be cognized if and only if it has moved, affected and stirred up an emotion or feeling in the observer such as its own nature requires. The perception of value is impossible unless the human behaviour is able to move the observer. Similarly, the observer cannot be moved unless he is trained to be affected, and unless he has empathy with the object of his experience. The subject's attitude toward the data studied determines the outcome of the study. This is why the humanistic studies of Western man and the social analyses of Western society by a Western scientist are necessarily "Western" and cannot serve as models for the study of Muslims or of their society.

Western social scientists impudently declare their investigations objective. But we know that they are biassed and that their conclusions are of limited significance. Dilthey's "sociology of knowledge" was not yet there to teach them that their presumed objectivity was a dream. Anthropology was the most daring of all, since its objects — the "primitive" societies of the non-Western world — were silent data, incapable of raising a critical finger at their masters. Theory after theory was erected to force the data into a mould, the categories of which were part and parcel of the Western world-view. The Western mind was still a long way from realizing, with the breakthrough of phenomenological axiology, that understanding the religions, civilizations and cultures of other peoples required an opposite bias, empathy with the data, if the data were to be understood at all. Although this discovery belongs to the comparativists and historians of religions and civilizations — the social scientists have hardly yet taken notice of it — we can say with certainty that the data of the social scientist, *viz*, the behavioural elements, are carriers of another element of a different nature — the valuational. Blindness or insensitivity to this distorts the investigation and vitiates its conclusions. Sympathy for it and openness to its moving power is a condition of its cognition and, therefore, a necessary complement if the conclusions are to be true to fact. The attitudes, feelings and hopes of individuals and groups do not speak out except to the sympathetic listener who welcomes being affected by, and thus emotes with them.

His attitude to them is crucial. Unless he is experienced and empathetic, they will escape him and thus vitiate the examination.

C. *Personalist Versus Ummatic Axiology*

The first argument of this section has shown that Western social science is incomplete; and the second, that it is necessarily Western and is hence useless as a model for the Muslim learner. The third will show that Western social science violates a crucial requirement of Islamic methodology.

Perhaps the most distinctive characteristic of Islamic methodology is the principle of the unity of truth. This principle holds that truth is a modality of God and is inseparable from Him, that truth is one just as God is one. Reality does not merely derive its existence from God Who is its Creator and ultimate cause; it derives its meaning and its values from His will which is its end and ultimate purpose. Its actuality has no meaning other than its fulfilment or non-fulfilment of value. Indeed, reality has become actual so that it may be an example of the divine will. It is therefore to be studied in the modality of value-realization or value-violation. As such, i.e. outside that modality, reality is nothing at all. It is hence invalid to seek to establish a knowledge of human reality without acknowledging what that reality ought to be. Any investigation of a human "is" must therefore include its standing as an "ought to be" within the realm of possibility.

This principle of Islamic methodology is not identical to the principle of the relevance of the spiritual. It adds to it something peculiarly Islamic, namely, the principle of ummatism. This principle holds that no value, hence, no imperative, is merely personal, pertinent to the individual alone. Neither value-perception nor value-realization pertains to consciousness in its personal moment, to its individual, secret relation with God. Islam affirms that God's commandment, or the moral imperative, is necessarily societary. It is essentially related to, and prevails only within, the social order of the *Ummah*. That is why Islam entertained no idea of personal morality or piety which it did not define in ummatist terms. Even *salāt*, the utterly personal encounter with and worship of God, Islam declared a means to the altruistic and other-related imperatives of morality. Indeed Islam made its religious value dependent upon them. That is why Islam prohibited monasticism and celibacy; transcribed its religious and ethical ideals into *sharī'ah*, or public law; and restricted its ethical precepts to public institutions which can thrive only if the state itself is Islamic. This is the significance of Islam's transcendence of the limits of Christian morality. Whereas Christianity defined salvation in terms of intention, i.e., the personal moment of consciousness, Islam defined it in terms of the act, i.e. public entry into the realm of space, time and society. In the former case, conscience was the ultimate tribunal on earth; in the latter, it consists of public law, public court, public sanctions, and rewards and punishment by God in history. The whole of eschatology Islam recast in a way to buttress this history-bound edifice of ideas, values, laws and institutions. Even Islamic knowledge itself, the knowledge of the will of God as given in revelation was made possible by Islam only as the subject of perpetual societistic effort on the part of society through *ijamā'* and *tawātur*. In the Islamic view, the axiological and ummatic are equivalent and convertible. Together, they constitute an intrinsic dimension, a *sine qua non* condition, of reality. There can be no knowledge of that reality without value, and there can be no religious or moral value except in the *Ummah*. The story of Hayy ibn Yaqzan illustrates this fact of Islamic consciousness dramatically. Having discovered Islam by natural reason and having, by his own effort, reached cognizance of its world-view and ethical vision, Hayy set forth on the open seas in search of the *Ummah* without which he could no longer exist.

The West has separated the humanities from the social sciences because of the considerations of methodology. This separation succeeded in banishing from the social sciences all valuations except those based on instrumental ends. "Scientific" objectivity could not tolerate them; and they were dumped in the humanities where concern and application of them became utterly personal and individualistic. This deliberate purge of the social disciplines from all considerations of ultimate value laid them open to whatever determinant happens to affect them. It proffered to the factual the power to constitute its own norms. The principle of the factuality and hence of the axiological autonomy of the social actualities being investigated led inevitably to the moral deterioration of society. Kinsey's value-free sex research diverted attention from adultery to the prevention of pregnancy. On the other hand, the assignment of the humanities by the West to a place outside science exempted them from the rigour of objectivity. By relegating them to a realm where scientific objectivity is not required, and by definition can never be achieved, it laid them bare to the attacks of relativism, scepticism and subjectivism. This helped to corrode their influence further and blight the power of their materials (faith, creed, hope, the good, duty, the beautiful, etc.) to determine life and history.

IV. *The Islamization of the Social Sciences*

1. All learning, whether it pertains to the individual or to the group, to man or to nature, to religion or to science, must reorder itself under the principle of *tawhīd*, i.e. that Allah (*subhānahu wa ta'ālā*) exists and is One, and that He is the Creator, the Master, Provider, Sustainer, the ultimate metaphysical cause, purpose and end of everything that is. All objective knowledge of the world is knowledge of His will, of His arrangement, of His wisdom. All human willing and striving is by His leave and permission. It ought to fulfil His command, the divine pattern He has revealed, if it is to earn for its subject happiness and felicity.

2. Pre-eminently, the sciences which study man and his relations with other humans ought to recognize man as standing in a realm dominated by God metaphysically as well as axiologically. These sciences include human history — the realm in which the higher levels

of the divine pattern are to be realized. Properly speaking, they ought to be concerned with the *Khilāfah* of God on earth, with man's vicegerency. And since man's vicegerency is necessarily social, the sciences that study it should properly be called Ummatic. Muslim learning repudiates the bifurcation, humanities/social sciences. It calls for reclassification of the disciplines into natural sciences dealing with nature, and ummatic sciences dealing with man and society. If, in the Association of Muslim Social Scientists we continue to call them social, we do so in defiance of the West which insists on separating them from the humanities. We must remember that the study of society cannot be free from judgment and valuation and is therefore subject to the same rigour, or absence of it, as philosophy, theology, law, literature and the arts. Conversely, the humanities are as much concerned with the *ummah* as the so-called social disciplines, and are capable of applying the same principles of validation to their materials and conclusions.

3. The Ummatic sciences should not be intimidated by the natural sciences. Their place in the total scheme of human knowledge is one and the same with the difference lying in the object of study, not the methodology. Both aim at discovering and understanding the divine pattern: the one in physical objects, the other in human affairs. Understanding the pattern in each realm certainly calls for different techniques and strategies; but as examples of the divine pattern the two are subject to the same laws of verification. Apart from giving details of rituals, describing the transcendent, and reporting about the unknown past, which may not be subject to critical verification, Islam has given us nothing by way of *naql* or tradition, which is not as confirmed or confirmable by reason and understanding, as what we have received or continue to discover by way of *'aql*, or reason.

4. The West claims that its social sciences are scientific because they are neutral; that they deliberately avoid human judgment and preference; that they treat the facts as facts and leave them to speak for themselves. This, we have seen, is a vain claim. For there is no theoretical perception of any fact without perception of its axiological nature and relations. Hence, instead of withholding analysis of the axiological aspect and allowing axiological considerations to determine the conclusion surreptitiously, a genuine scientist will seek to analyse *all* the aspects of a given phenomenon, and will do so in the open. He will never claim to be objective when he in fact is prejudiced, to be complete and thorough when he is in fact reductionist, to talk about

human society when he is in fact referring to Western society, of religion when he is in fact referring to Christianity, or of social and economic laws when he is in fact referring to common practices of some Western societies.

5. Finally, Islamization of the social sciences must endeavour to show the relation of the reality studied to that aspect or part of the divine pattern pertinent to it. Since the divine pattern is the norm reality ought to actualize, the analysis of what is should never lose sight of what ought to be. Moreover, the divine pattern is not only normative, enjoying a heavenly modality of existence removed from actuality. It is also real in the sense that Allah has predisposed reality to embody it, a kind of *fitrah* existence which Allah, in His mercy, implanted in human nature, in the human individual or group, in the *Ummah* as an ongoing stream of being, which moral action pulls out into actuality and history. Every scientific analysis would therefore endeavour, if it is Islamic, to expose this immanent divine pattern in human affairs, to underline that part of it which is *in actu*, and that part of it which is *in potentia*; to expose the factors realizing or impeding the completion of the process of embodiment, to focus the light of understanding upon the relations of that process to all the other processes of ummatic life.

The Islamic social scientist is endowed with the cause of Islam. The divine pattern in human affairs is the object of his constant attention and care as well as of his hope and yearning. He is not only scientific in the sense of not leaving out the axiological aspects, but is preeminently critical of reality in light of the divine pattern.

Per contra, the Western social scientist cannot afford to be critical of the ultimate purposes or ends of society but only of the means thereto, because of his conscious commitment to description rather than to advocacy. This commitment, however, was hardly ever honoured because his denial of the relevance of the spiritual was, in the final analysis, a denial of relevance to Church-related values, and an assertion of relevance to utilitarian and cultural values which he honoured subjectively, in tribal or Protagorean fashion. The Islamic social scientist, on the other hand, has maintained an open and public commitment to the values of Islam, an ideology which lays a rational, critical claim to the truth. He is not afraid or ashamed of being corrected by his Muslim or non-Muslim peers; for the truth, in his view, is none other than the intelligent reading of nature in scientific report and experiment, or the reading of God's revelation in His holy book. God is the Author of both; and both of His works are public, appealing to no authority other than that of reason and understanding.

From such a view, the Islamic social scientist is capable of bringing a new critique to social science. Loyalty to means and instrumental ends has caused Western social science to degenerate into "strategic studies", i.e., studies of strategies leading to goals the validity of which nobody claims to be critically establishable. Through commitment to Islam, the social scientist is bound to regard man as God's vicegerent, whose duty is to actualize value in history. Islamic social science can therefore humanize the discipline and reinstate the *humanitas* ideal in the life of man, the being whom Western social science has taught to regard as a helpless puppet of blind forces.

V. Recommendations for Action

A. *Human Resources*

Every Muslim university around the world is understaffed, and no Muslim university, not one, can claim that its social science curriculum is Islamic. That there may be a few brighter spots around the Muslim World is not denied. But the undeniable and most crying need of Islamic education is for human resources. There are hundreds of thousands of M.A.'s and Ph.D.'s, but few among them are even aware of the problem of Islamizing the disciplines; and there are legions whose brain-washing by the West has been so complete as to make them committed enemies of Islamization, or at best, lethargic, indifferent, even cynical observers of the scene, non-moving and immovable. The following four measures should be taken to stir us from our present lack of action:

1. The formation of an association of Islam-committed scientists, whose purpose is to spread and intensify an awareness that the problem exists, that the problem is extremely grave and dangerous, and that all efforts at reconstructing the *Ummah* will be vain unless our intelligentsia become aware of the *Ummah*'s Islamic mission and translate that mission into directives for living in the various fields of human endeavour (formed by the few Islam-committed social scientists aware of the problem, who are active). Fortunately, such an

association exists, and it is more than six years old. But if it did not, the first step would be to bring it into being.

2. Such an association ought to relate to one or more Muslim universities, which would sustain it with human and other resources, and provide classrooms and lecture halls as a laboratory for the association's discoveries and an arena for its accomplisments.

3. The first measure such an association ought to take is to locate and identify its potential members. These are, first, "doctoral graduates" in the disciplines whose commitment to Islam has urged them to seek the Islamic relevance of their knowledge; second, *'ulamā'* in the traditional sense whose reasonableness, sense of history, alarm at the disintegration of Islamic knowledge or *angst* at the forceful corrosion of Muslim society, have urged them to add to their command of the Islamic tradition the tradition of Western learning; thirdly, general Muslim talent for creative ummatic thought outside academic ranks.

4. The vision of Islamizing the social sciences is shared only by a few scholars and is not readily available to anyone. Training Muslim talents in the vision is the next foremost duty. Such training must be carried out on the post-doctoral level. The association should arrange for its members to complement their education in programs especially designed to suit each member's needs. Such programs should consist of intensive courses, intensive seminars, reading and research projects, as well as conferences.

B. *Materials of Study and Tools of Research*

Annotated, topically arranged bibliographies concentrating on each discipline, and each major problem in the discipline, ought to be prepared for both the Islamic and Western traditions of learning. The former is rich; but its riches lie hidden under titles unrecognizable by the systematic categories under which the modern mind usually operates. Specialists in each section or field of Islamic literature would have to locate and identify the relevant passages, and discipline experts to determine the relevance of such items in the tradition. The Western tradition in each discipline would have to be researched for materials pertinent to the issues the *ummah* faces at present. Beyond these bibliographical surveys, topically systematized anthologies of readings should be prepared for each discipline — indeed, for each problem or area within the discipline.

Analytical surveys or articles dealing with the historical development of the problem or discipline, or with the contemporary state of research, ought to be prepared by the experts for use by the less advanced in the field. Assuming the urgency of the need, this is the speediest way to put at the disposal of the creative mind the tools necessary for extending the frontiers of Islamic knowledge.

C. *Creative Works*

With the human resources mobilized and trained, and the tools of research prepared and assembled, special programmes of workshops and seminars ought to be designed to the end of enabling the available talents to think out, and present for ready use of the understanding, creative articles, essays and books establishing the relevance of Islam to the various disciplines and to the major problems within each discipline. This programme should be built around, or lead to, individual or group research and writing assignments.

It is only after we have built some tradition of creative Islamic thought in the sciences that we may incept the preparation of textbooks for use by our educational institutions. The textbook is of little use without the trained teacher, or with a teacher whose loyalties are elsewhere, and whose vision and knowledge are deficient.

All three steps can be telescoped into one process, because all three are mutually complementary.

Chapter Two
Sociology and Muslim Social Realities

Ilyas Ba-Yunus

Ilyas Ba-Yunus was born in Pakistan in 1932. Professor of Sociology, State University of New York, Cortland, New York 13045., (since 1973). Previously: Assistant Professor of Sociology, Bradley University Peoria, Illinois 61606, U.S.A. (1968–1973), worked in Oklahoma State University as instructor in Sociology (1966–67), and as instructor in Geography at Winona State College, Winona, Minnesota, U.S.A. Fellow (by invitation) of the Inter-University Seminar on Armed Forces and Society. Publications include "Islamic Personality," *Al-Ittihad* (Summer, 1968); *Muslims in North America; Problems and Prospects* (M.S.A. of the U.S. and Canada); "Ethnic Communities in the Melting Pot" (Proceedings of the Fourth Annual Conference of the Association of Muslim Social Scientists); "Jamaican Boys in England: A study in Conformity and Delinquency" (*Selected Papers*, 11th Annual Convention of the Royal Society for Delinquency Prevention, England). "The West Midland Teddies: An Empirical Test of the American Hypotheses" (*International Review of Modern Sociology*, Fall, 1974); Chief Editor, *The Third World Review*.

Sociology as a discipline of inquiry in human interaction, covers a wide range of topics. Sociologists have concerned themselves with conflict and consensus, competition and cooperation, organization and disorganization, deviance and conformity, order and changes and other processes which fall in the scope of human interaction. On the one hand, they have focused on interpersonal relationships in small group situations. On the other hand, they have been concerned with processes of much larger magnitude which occur within or among societies. Polity, economy, birth, death, family, migration, education, law, justice, religion, crime, recreation—whatever human beings do in relation with other human beings has been considered as a bonafide area of sociological analysis.

Its breadth and scope notwithstanding, contemporary sociology, as Marsh pointed out (1967:19) "has been developed in a small corner of the world and may, therefore, be highly limited as a universal scheme." This is one of the more serious dilemmas of modern sociology which sociologists can ill afford to hide: not only that Western thought has remained distant from and often in conflict with local perceptions in non-Western societies, but that specific sociological theories do not explain many a problem which these societies are facing today.

Examples of this shortcoming in sociology are many. Theories of crime and delinquency which are based on experiences and researches in the inner parts of the cities like Chicago and New York, do not explain crime in the Soviet Union (Connor, 1972), in Pakistan (Ba-Yunus, 1976), in Egypt (Hassan, 1977), in Indonesia (Bannister, 1973) and in other societies. Modern sociology has explained religion on the basis of experiences in the organized Christian church and the sectarian dilemmas in Christianity. Naturally these views cannot explain religious experiences of non-Christian societies. Major theories of social change are based on the aftermath of industrialization and modernization of Europe and America (for instance, Moore, 1973 and 1977; Inkles, 1969; Inkles and Smith, 1974). These and similar views have come under severe attack for being ideologically biassed (Hechter, 1975; Wallerstein, 1975; Hill, 1975). At the same time, Portes, a major contributor to modernization research, has attacked these theories for promoting counterproductive models of economic development (1973).

There are several other theories dealing with many other social phenomena which can be and have been criticized in this respect. Whether they are dealing with crime, social change, inequality or religion, existing sociological theories seem to be inadequate in explaining non-Western societies. This shows the culture-bound character of contemporary sociology; and if we look closer we would find gross differences in approaches among Western sociologists from one country to another. There have been notable attempts to relate sociological developments with economic, cultural, and ideological trends in America (Mannheim, 1953; Gouldner, 1970), Germany (Mannheim, 1953; Aron, 1964), France (Cuvilliar, 1958), Britain (MacRea, 1958), Italy (Direnzo, 1972) and other Western countries. Even if these sociological differences from one Western country to another disappeared because of closer interaction among these sociologists, we are still left with an awkward fact that Western sociological approaches are based on assumptions and research findings which are foreign to social realities in non-Western societies.

And when one shifts one's focus from non-Western societies in general to Muslim societies or the Islamic culture area in particular, one finds that the systematic study of Islam is a grossly neglected field in sociology. There are hardly any sociological studies of Islam and Muslim societies. For instance, Turner writes:

> An examination of any sociology of religion textbook published in the last fifty years will show the recurrent and depressing fact that sociologists are either not interested in Islam or have nothing to contribute to Islamic scholarship... There is no major tradition of sociology of Islam and modern research and publications on Islamic issues are minimal. Most academic sociologists who are responsible for teaching sociology of religion courses in universities will steer consciously or unconsciously away from an analysis of Islam simply through lack of basic teaching sources... There is consequently a need for studies of Islam which will raise important issues in Islamic history and social structure within a broad sociological framework which is relevant to contemporary theoretical issues (1974: 1–2).

Not only have Western sociologists in general ignored Islam as a unit of analysis, but also on the rare occasions that they have done so, they have remained inconsistent in their approach to Islam. For instance, Turner writes about no less a sociologist than Max Weber:

> The importance of Max Weber for modern sociology does not depend solely on the contributions he made to sociological knowledge through his substantive studies of India, China and Europe. These studies are important, but Weber also made a massive contribution to contemporary sociology by outlining a special philosophy of social science and a related methodology which attempt to present the actors' constitution of social reality by subjective interpretation... In my opening chapters, my argument will be that, in his observation on Islam and Muhammad Weber was one of the first sociologists to abandon his own philosophical guidelines. It follows that my attitude toward Weber is genuinely ambiguous. On the one hand, Weber does provide a stimulating framework within which one can raise important theoretical issues in relation to Islamic development. On the other hand, Weber inconsistently applied principles which he declared were crucial to an adequate sociological approach (1974:3).

The above observations show that modern sociology is inadequate for the analysis of non-Western societies in general. In the case of Islam, however, it is not merely inadequate; it may be misleading. It follows that for a correct understanding of Islamic ideology and Muslim societies we need an Islamic sociology. Without this development, the sociological view of Islam will continue to remain distorted. My purpose in this paper, therefore, is to present a frame of reference which would reflect an Islamic orientation as opposed to Western orientations—whether Marxian or capitalistic. In doing this, first I shall discuss some of the more dominant trends in contemporary sociology. Secondly, I shall discuss more salient aspects of Islamic ideology in order to show how Islamic sociology differs from Western sociology. Finally, I shall present a model for Islamic sociological analysis.

Contemporary Trends in Sociology

Although Ibn Khaldun introduced his science of society (*'Ilm al 'Imran*) in and around the year 1377, it is customary to trace the roots of modern sociology in the writings of a French philosopher, Auguste Comte (1798–1857), who was born almost four hundred and fifty years after Ibn Khaldun. Whether or not neglect of Ibn Khaldun reflects an ethnocentric bias on the part of Western sociologists, it certainly substantiates our view that the modern sociologist is quite ignorant about Islam and the Muslim World.[1]

Although Comte is hardly a source of much inspiration for modern sociologists, yet his emphasis on positive methodology in studying the human phenomenon as distinct from speculative social and political philosophies of the past still constitutes a fundamental belief in modern sociological creed. Like their counterparts in physical and natural sciences, sociologists are supposed to study their subjects — societies, groups, or individuals — with a sense of detached empathy. Recognizing that their subjects are human beings, sociologists caution themselves against ethnocentrism or contamination of their views with their own personal or group values. Sociology being a science, sociologists are expected to base their views on empirical findings alone. What is not empirical is supposed to be speculative and, therefore, not necessarily valid.

Despite these warnings which are essential parts of any introductory text book in sociology today, most sociological views reflect value biasses of their authors and most sociology is quite speculative in the final analysis. It is perhaps quite possible to exert some control over such biasses, but the degree of control always varies from one sociologist to another. Although sociologists have presented a number of views about human society, in the following I shall restrict myself to three main approaches which are in vogue among sociologists today:

I. The Structural Functional approach

I. The Structural Functional approach which came into prominence during the late Thirties, has a macroscopic view of society.[2] It was first introduced by Talcott Parsons at Harvard (1937) and was promoted by a whole generation of his students and their students mainly in America. This approach is not unique to sociology. In fact, the basic argument of this view of society has been borrowed from biology.

There are two fundamental assumptions on which this approach is based. The first assumption is that society is a system which is made up of interdependent sub-structures in such a way that change in one part is automatically reflected in change in other parts as well. The task of sociological analysis, therefore, is to find out what is influencing what. The second assumption is that any established structure or activity, however injurious it may look to an outsider, has a function of maintaining other activities or structures in a social system. Examples of these structures in society as a system are family, economy, polity, religion, education, recreation, law — just to name a few. Every structure is supposed to be maintained by the roles which people play in their individual status within these structures. Finally, these roles cannot be appropriated except by learning the rules which are developed as a result of a general consensus in society. This is how this approach has also come to be known as the *consensus model* whereby people are seen as cooperating, agreeing, sharing in making rules so that society assumes the structure of an ongoing system.

This approach has been criticized on the grounds of ignoring the role of conflict, revolution and disputes which cannot be ignored in societal analysis. More than that, this approach has also been criticized for advocating the status quo (what is there is good) in order to maintain the capitalist structure of Western democracies in general and that of America in particular. In fact, this approach equates modernization with Westernization i.e., industrial institutions cannot be developed and maintained without a quick emergence of institutions which characterize Western societies, for instance, materialism, secularism, democracy and a devotion to work.

I would hardly criticize Parsons and his students for keeping to modern capitalistic society as a model for their approach in sociology. However, I would question the uncritical application of this model to non-Western societies in general and to Muslim societies in particular, societies which have had histories of uprisings, military coups, colonialism and other external influences arising from travel, education, and trade.

II. The Marxian or Conflict approach

II. The Marxian or Conflict approach provides the most important alternative to the structural functional approach in macro-sociology today. Karl Marx (1818–1883) is most famous as the originator of the international socialist movement. While a large part of his writings was devoted to the propagation of this movement, many of his doctrines are recognized, in a modern sense, as sociological.[3] However, whereas an ideological bias is only implicit in the writings of structural functionalists, sociological followers of Marxian doctrines use his sociological as well as ideological guidelines quite explicitly.

Marx's sociology is based on two fundamental assumptions. He viewed economic activity as the basic determinant of all social activity; and he viewed human society mainly in terms of conflict throughout history. Economic activity in society dominates all other structures in society like political organization, family, religion, law, art, literature, science and morality. He saw the mode of economic production throughout human history to be such that most of the economic resources were in the hands of a select few in the society while the rest of the society was condemned to work for them and remained at the mercy of their goodwill. Thus Marx saw human society divided in two classes, that of owners who always exploited and that of workers who were exploited. This continuous exploitation, according to him, necessitated revolutions. However, according to Marx, in the absence of any other mode of economy, the leaders of revolution followed the same system of exploitation by themselves controlling most of the resources and again condemning the masses to the status of workers. This process, according to Marx, was further intensified in the wake of industrialization in Europe.

Basing his view of human society on this process of history, Marx presented his ideology of socialism – a final solution whereby all the resources would be owned by everyone through legitimate representatives in the government. Because, ideally there would not be any individual owners left, according to Marx, there would not be any further strife, conflict and exploitation.

As mentioned above, followers of Marx remained directly or indirectly tied to his philosophy as well as to his ideology. On the one hand, human society has been seen as being full of conflict and disputes instead of consensus and cooperation. Law is seen as being made by the exploiters. Religion is seen as being promoted by the exploiters for their own sake. Crime is seen as a necessary consequence of exploitation as well as being a result of exploitative legislation. In short, exploitation and conflict are seen as the basic processes of human society as an ongoing system. Marx's model of human society is, therefore, known as the conflict model.

On the other hand, followers of this model do not seem to hide their conviction that socialism is the only remedy for this exploitation and perpetual conflict in society. Until recently, sociologists in socialist countries were almost the only ones who followed this model along with a few others in Western European countries. However, anti-Marxian sentiments are declining in America and a considerable number of American sociologists are presently beginning to adopt the Marxian model of society — and his ideology as well.

One does not have to be a follower of Marx in order to see the existence of economic exploitation within and among societies. However, it is equally apparent that all human activity is not determined by the mode of production as hypothesized by Marx. Neither does it seem true that societal institutions emerge and are sustained as a result of inter-group conflict. Further, human societies are found to be much more complex in their class and caste structures than the bi-modal structure seen by Marx.

It is often claimed that the conflict theory is better suited to account for changes in societies and that functionalism is better able to explain the persistence of various features of society. In any case, it is probably a safe generalization that conflict theorists give greater emphasis to disputes and social change because they are personally more likely to desire radical social change. Functionalists give more emphasis to stability because they are personally less likely to desire radical change and more likely to prefer reform of present arrangements. Thus the ideological thrust of their theories reflects a desire on the one hand to know how to pull societies apart and on the other hand to know how to prevent societies from being pulled apart. It is much more plausible to assume that societies experience both consensus and conflict internally and internationally; that societies may change because of internal convulsions as well as external pressures, but that they may also change because of slow and gradual change in morals, habits, fashions and technology, through consensus and mutual sharing.

These two macro-perspectives still do not tell us about the nature of interaction among or between people in limited situations. They by-pass human rationality and decision-making as if human beings are like billiard balls which are tossed around by forces beyond their own control. Sociology would be perhaps incomplete without an approach which looks closely at human interaction, which is the building block of human society. There is no doubt that often knowingly or unknowingly human beings generate larger processes which force them in some specific directions. However, how can sociology ignore the existence of the social individual who perceives these processes, gives them meaning, resists them or decides to go along with them? Symbolic interactionism or self theory is such a micro-perspective in sociology that it may be quite speculative at the present stage of analysis, but it is the least guilty of ideological bias; although, as we shall see later, this approach is considerably affected by the social conditions in which it is nurtured.

III. The Symbolic interactionism or Self theory

III. The Symbolic interactionism or Self theory, more often known simply as the *interactionist approach*, as the name implies, begins its analysis with social interaction at the most minimum level. From this micro-perspective, unlike other brands of social psychology, it hopes to broaden its scope of analysis in order to capture the whole of society as a multiple interaction process. Human beings are seen as learning situations which may be conforming or full of deviance, situations of economic and political transactions, situations within or outside of the family, situations of play and education, situations of formal or informal organization, etc. It is on the basis of this learning process that individuals are seen as defining or interpreting further situations that they find themselves in — directly or indirectly, physically or psychologically. It is on the basis of these interpretations that they develop justifications in order to make decisions to act or not act. That these justifications may often be wrong and even not acceptable by others, is fundamentally assumed. However, this approach does not assign causation to these justifications. If causation is attributed to anything it is to the actor himself. Every actor is supposed to be the builder of and responsible for his own act. Even when a person is seen as being unable to challenge circumstances, this phenomenon is supposed to be explained only from the perspective of the actor himself. Only habitual or reflex actions are supposed to be exempt from this logic.

The roots of the interactionist approach are embedded deep into the rationalism of John Locke and the idealist epistemology of Kant. However, it emerged as a distinct field within sociology as a result of the contributions made by John Horton Cooley at the University of Michigan, Robert Park, William I. Thomas and most importantly, George Herbert Mead, all of them at the University of Chicago. Mead taught at this university from 1897 to 1933. Without any traceable bias, however, this theory was extremely speculative from the very beginning. Although Thomas and others did emphasize empirical research and produced their own monumental works, the main ideas of this approach revolved around Mead's hypotheses which remained unverified all through his life. His examples remained mostly hypothetical, unauthenticated and common-sensical rather than research findings. He taught for more than thirty years; but he never wrote a book. The very first formal statement of his approach appeared when his students compared their lecture notes and published a book, *Mind, Self and Society*, under his name, one year after he died.[4] Although because of a rather fast spread of structural functionalism in America about the same time the interactionist approach remained relatively subdued, yet a few staunch adherents of this approach continued to refine its theories and developed their own research strategies and data. As a general disenchantment with the structural functional approach is setting in in America these days, the interactionist approach is becoming more prominent.

The symbolic interactionist approach is described as being non-scientific (not unscientific) in that it assumes unpredictability of human action while trying to give an understanding of it from the perspective of the actor himself. Thus, trying to understand human society on the basis of how people see it and how they made their decisions, this micro approach claims to have the capability of magnifying its focus from less complex to more complex processes in human affairs.

Sociological Doctrines of Islam

As in the conflict approach, Islam gives a dialectical picture of human society through historical evolution. From this point of view, human history describes a conflict between those who brought the Divine guidance in order to establish a just society and those who always opposed these prophets. As Prophets were able to establish a just order, their descendants would soon forget and even in time contaminate the true message. Hence the need for another prophet who would again try to establish a new order congruent with the cultural evolution of human society over a period of time. Finally, at a certain juncture in the cultural development of man came the last and the most comprehensive formula of societal life — Islam through Prophet Muhammad. Thus, like the Marxian view, the Islamic view of human society is ideologically committed.

The systemic view of human society which characterizes the structural functional approach is nothing new to an ideological Muslim. For him, not only human society, but the whole universe is a system. For him, Islam came to establish a well integrated system functioning under the rules provided by God. Any society which deviates from this ideal is a society in conflict which in time creates disintegration. And if we look closer, it may not be difficult to find the indeterministic emphasis which has become the trade mark of the symbolic interactionist approach in sociology, as the basis of law and punishment in Islam. Denying the very idea that human beings carry the burden of "original sin", Islam describes Adam and Eve and their progeny as being capable of conforming to, as well as deviating from, the Divine law out of their own power of decision.

Emanating from the Divine, Islam is the natural law of human interaction. There is no other just way of human interaction. God created the universe and provided laws of structure and change in it. Because all objects in the universe, physical as well as biological, function according to the Divine law, they function harmoniously. Physicists, chemists and biologists try to discover these laws. God also created man and gave him the power of decision and rational thinking. Because man could err in decision making through his lack of knowledge, He also gave man the law of interaction which creates harmony and peace and removes conflict and exploitation. Without His law the universe would collapse. By not following His law, human beings would head toward mutual destruction. Islam, then, is not merely a formula of rituals. It is the process of obedience to the rule of God in human relationships in all aspects — economic, political, family, law, punishment, war, recreation, innovation, education and socialization. Emphasis is on deliberation. Any unintentional deviance from these rules is excusable. Any intentional deviance from these rules is not only punishable by society and in the hereafter, it sows the seeds of self destruction in this very world. Finally, since Islam is the natural law of interaction even if a Muslim society does not follow it, this society is automatically heading towards injustice and decline. Even if a non-Muslim society follows this law, it has to prosper.[5]

But, what is Islamic ideology? In these days, modern intellectuals have come to see the world as being divided in two ideological positions — socialist and capitalist. However, it is too simplistic to reduce the whole world to only two ideological practices. It is much more logical to see the world according to a continuum with capitalism at one end and socialism at the other. In between, there may be different shades of capitalism and different shades of socialism. Islam rests midway on this scale so that from the capitalist end it may look like socialism and from the socialist end it may look like capitalism. The truth, however, is that it is neither. It was given to mankind long before either capitalism or socialism came into existence as recognized ideologies. It avoids the extremes both of capitalism and of socialism. Above all it is not a man-made ideology. It emerged directly from the Divine will. Logically, therefore, it is unlike any other human ideology although for the sake of comparison one may apply the above-mentioned scale.

In its political form, democracy is the only structure which approximates to Islam. God is the head of the state. It is the responsibility of the believers to elect or select decision maker(s) from among themselves and obey his authority as long as he obeys the authority of God as given by the Prophet. This describes the process of *Shura* which is also applicable to all affairs in which there is a dispute or disagreement. However, *Shura* must not be confused with modern democracy whereby people seek office in order mainly to enhance their own power. *Shura* in principle denies authority to a power seeker. *Shura* is mainly in the service of God, very much like the election of the Pope. However, unlike the Pope the elected *Ameer* of the Islamic state has power over all societal issues. His power is not limited to theological issues only. Roughly translated as consultation, *Shura* is also the basis of legislation and change in society, so much so that even the *Ameer* may not be beyond the reach of the *Shura*. A direct consequence of this principle is the institution of justice. The judge or the *Qadi* has to function directly in obedience to God and His Prophet and the legislation which has come down to him through the process of *Shura*. The judge or the *Qadi*, is, thus, autonomous, and all other decision makers and the elites have to obey his court-room decisions.

In its economic structure, Islam has allowed free trade and the ownership of individual property. However, it has emphatically prohibited interest which is at the root of modern banking practices. Islam has prohibited gambling and it has levied *Zakat*, the poor tax, which is supposed to be the right of the needy and not an act of charity on the part of the more well-to-do. Further, Islam has instituted rules of inheritance which divert a substantial portion of the property of the deceased to the needy and the poor. These are the minimum requirements of the Islamic economic system. Beyond these minimum requirements, there are strong warnings against those who hoard, take undue profit and those who love their property more than working in accordance with Islamic justice. Thus, on the one hand, Islam inhibits the undue accumulation of resources by a small number of people. On the other hand, it allows a free market economy.

Beyond these rules for macro-processes in society, Islam has provided rules for interpersonal and personal conduct such as marriage, divorce, manners of speech and a general posture toward others. This whole ideology is, however, supposed to be maintained as a result of a personal commitment to God. This commitment is made and enhanced by observing prayers five times a day and fasting in the month of Ramadan, not in solitude but as a public activity so that an environment of piety in the community of the believers may develop.

Limitations of space do not allow me to go beyond the most essential features of Islamic ideology outlined above. Briefly, Islamic ideology focuses on both the micro and macro processes of human society so that the personal worship of God is not supposed to be an end in itself. Its end product is the establishment of the Islamic state, which others are invited to join.

The Islamic Sociological Approach

With the above remarks in mind, we may ask the question: how should a Muslim sociologist proceed? God has created man with the power of learning and of decision making. He has also provided man with an ideal system of interaction. Thus, from the Islamic point of view there is nothing wrong in assuming conflict and consensus in human affairs. However, human society cannot be considered only in terms of conflict or of consensus. Rather, conditions of conflict and consensus may be assumed to be present to the degree that society is removed from or is approaching the Islamic ideal.

There are at least two major points on which the Islamic sociological approach would differ from contemporary sociology. The first concerns the general treatment of religion by sociologists. Structural functionalism and conflict theories much more explicitly, and symbolic interactionism only implicitly and perhaps ambiguously, have assumed that religion is one of the many things which happen in society. Following their experience in predominantly Christian societies, structural functionalists have treated religion as one of the institutions of society. Dürkheim, one of the foremost originators of functionalism, reduced religion almost to a totem pole — a cultural artifact — which once socially created, however, is supposed to provide a communal unity. This general assumption of Western sociologists is in conformity with modern capitalist philosophy i.e., religion and state are two separate things engaged in symbiotic relationships.

On the other hand, the conflict theory differs from the structural functional approach in looking at religion as necessarily bad, an "opiate for the masses" as put by Karl Marx. Otherwise, like their counterparts in the structural functional approach, conflict theorists have looked at religion as one of the institutions of society — a social artifact which, irrespective of its origin, serves as a tool in the hands of the exploiters in order to promote their hold on the poor and the weak. Contrary to what many other people believe, the conflict theory holds religion responsible for exploitation, supporting the oppressor, and generating crime and delinquency in society.

Symbolic interactionism, on the other hand, because of its micro focus does not directly concern itself with issues of a broader nature in society and is, therefore, non-committal in this regard. Adherents of this approach tend to focus on such issues as individual religiosity, the process of religious conversion, and the self-perceptions of people with regard to their respective religions. However, because most symbolic interactionists are Americans, their studies of religion have not been in any respect different from the way religion is treated in America — i.e. as one of the many forms of social interaction. This is brought about quite clearly in cases of conflicting situations in secular society. For instance, Burchard (1954) pointed out the dilemma of the military chaplain whose religious training demands of him love and mercy but who has to condone the killing of enemy soldiers. Likewise, Komorovsky (1953) pointed out the dilemma of Catholic girls in college as they faced a rather permissive environment. Although symbolic interactionists do not claim to generalize from their findings in respect of all societal situations, yet should they choose to do so, they would not seem to have any other alternative.

The Islamic approach would certainly have to differ from this treatment of religion in Western sociology. For Muslims, Islam came for the benefit of mankind; and as mentioned above, it is supposed to be the basic and all-encompassing force in society for institution building. Muslim societies, even those which profess a separation of religion and the state, experience religion, however incompletely, in all aspects of the lives of their people — nationally and internationally. Following the Western model, Muslim sociologists may not be able to analyse Islam as an ideology and may face severe difficulties in analysing Muslim societies as well.

Following Islamic ideological assumptions, sociologists may be interested in knowing the degree of divergence between the actual and the ideal. Should this be the main interest of a Muslim sociologist, as it must, he would have to develop an Islamic model, an "ideal type", against which all Muslim societies as well as Muslim minorities could be judged as to their degree of congruence to Islamic ideology. All kinds of measures — historical, statistical, case studies, participant observation, even experimentation — could be applied in using this model. The aim should be to generate comparative data in order to see how far Muslims are removed from Islam today.

Further, a Muslim sociologist may question the general assumption of western sociology regarding the role of religion in general. How far is it true that religion is only one of the institutions in other societies? Could it not be true that even in non-Muslim societies, Christian societies included, religion plays a much more important role than has been recognized so far? Quite a number of examples challenging this assumption may readily be cited. For instance, the State of Israel is based on religion (something which may turn a Muslim ideologist green with envy). In nearly all Western and other Christian societies, the oath of a political office or that of a witness in the courts of law is administered in the name of God (or on the Bible). A "reborn Christian" not only became the President of America, he has been successful in shattering the myth of liberalism in the country. There are many Western countries where religion has the status of the official faith of the state. There must be some reason why evangelism thrives in America. True, Christianity and other religions do not provide their followers with the rules of a complete social life, but a great majority of them do not necessarily seem to limit their religious self-perceptions to the confines of churches and temples only. Of course, a structural functionalist may argue that the influence of religion as an institution would be felt in all other institutions of society. However, the degree and the direction of this influence must be assessed by empirical research.

So far, Western sociological research has been guided by assumptions that reduce religion to the status of one of the many institutions in society. Because of these assumptions, research has never been allowed to look into the extra-institutional and wider effects of religion in society. The result has been the development of the sociology of religion, which deals with different correlations of religion in theory and practice. What is needed is a research strategy which would challenge rather than follow these presently held assumptions regarding religion and society.

As an extension of their approach, adherents of Islamic sociology must initiate such research. Models of religious effects should be developed as ideal types for societies to be studied in order to measure the degree of congruence of these societies to their own religions.

In short, Islamic sociologists have to be comparative sociologists in intra- as well as inter-societal analyses with an explicit religious rather than a purely materialistic and secular outlook toward human society.

The second major point on which Islamic sociology would be at odds with dominant trends in contemporary sociology would be in terms of what one might call applied sociology. Or, as Lynd (1939) put it: Knowledge for What?

With the exception of conflict theorists most of whom have not yet gone beyond merely championing the socialistic model of society, sociologists have a kind of identity crisis with respect to this particular issue. The question of the application of sociological knowledge necessarily raises the issue of value involvement. Most American sociologists have preferred to remain value-neutral i.e., they see their task as technical and themselves as technicians or scientific investigators who need not be concerned with the values that generated their studies or with the societal implications of the results. Notable exceptions to this position in America have been C. Wright Mills (1959) and Howard Becker (1967).

Many sociologists are engaged in tasks other than merely teaching and research. They seek and are given employment in different capacities in industry, business, and policy planning. In these capacities they may use their knowledge in order to solve problems of drug addiction, crime, delinquency, urban congestion, city planning, family planning, and other areas. In such capacities, sociologists have served not only as social analysts but also as social policy planners and decision makers. However, how these decisions are made or how they try to motivate people to initiate or give up some form of action is never made a part of sociology proper. Consequently, these applied strategies never become part of the academic training of new generations of sociologists in colleges.

Few societies in the world would continue to bear the cost of an academic discipline which does not prepare its students for its application. American college graduates with an undergraduate major in sociology are now beginning to feel this inadequacy of their discipline. Anthropologists were quick to understand this problem during World War II when they developed the so-called "survival kit" however dubious its value for the American soldier fighting in far-off and strange countries. Sociologists were also employed by the American military during and after World War II. "The American Soldier" series which involved such sociologists of repute as Samuel Stauffer, Robert K. Merton and many others, helped us understand the reasons for the superior performance of the German and the Japanese soldier relative to his American counterpart. Likewise, sociologists were given huge grants to study the problems of the American soldier in later wars in Korea and Vietnam. These sociologists, however, did not serve to improve the morale of the soldier and they could hardly help in developing war strategies. What came out of these studies is nonetheless academically valuable. The so-called reference group theory is to a very great extent an outcome of these World War II series in sociology. Subsequent grants were so lavish and so many sociologists benefited from them that sociology of the military is now a small but fast-growing field within sociology.

Generally it has not made much difference whether a person is a structural functionalist or a symbolic interactionist. Emphasis in research has always been on enhancing theoretical knowledge and not on analysing or developing strategies which could manipulate this knowledge in order to solve practical problems. Indeed, if Comte were living today, he would certainly disown his successors. Or, may be, they would disown him.

The Islamic sociologist, on the other hand, has to be value-involved by the very logic of his approach. As a comparative sociologist, he may find discrepancies between existing social processes of both a macro and a micro nature and the Islamic ideals. However, his job would remain incomplete if he did not pay attention to the practical strategies as to how this gap could be minimized both at macro and at micro levels. In short, unlike any other sociologist in the world, the Islamic sociologist has to play the role of an analyst, a critic, and at the same time a strategy maker; and this analysis, criticism and strategy planning must become an integral part of the Islamic sociological approach.

Summary and Conclusion

My major contention in this paper has been that contemporary sociology as it is taught in Western universities may not only be inadequate, it may also be misleading in understanding Islam and Muslim societies. After a brief review of major sociological approaches which are in vogue today, I have stressed that a new approach — Islamic theory — is needed not only to analyse Muslim societies but also to understand the ideological role that religion plays in human societies in general. I do not have any preference for one research technique over any other — historical analysis, survey research, case studies, participant observation, and experimentation. The general methodology of research in Islamic theory, however, has to be comparative in its general design so that the degree of divergence of Muslim societies from the Islamic type can be measured.

In its substance, Islamic theory has to be critical. I cannot envision that a follower of this approach would be value-neutral. The Islamic theorist has to be critical, i.e. one who discusses the degree of divergence between society and its ideology from a historical perspective, with the help of statistical correlations, as a result of participant observation or in terms of any other research technique applied.

Further, the role of Islamic theory would remain incomplete unless future plans were envisaged on the basis of the above findings. The Islamic theorist, then, has to assume the role of a planner of strategies so as to assess which aspects of Muslim society need to be drawn toward the Islamic ideal.

To summarize then, Islamic theory has to be *comparative*. It has to be *critical*; and it has to be *strategic* i.e., aimed at planning for the future. Of course, all adherents of this approach are not expected to specialize in all these aspects. However, specialization in one area must not be an excuse for ignoring other areas. In fact, all three aspects of Islamic theory must be introduced as essential aspects of the sociological curriculum at high school, undergraduate and graduate levels.

By its very nature, this approach has to draw most of its support from sociologists and other educational planners in Muslim countries. Equally strong support, however, has to be initiated by government and private sources in business and industry. Without sizeable grants for research, library facilities, original writings, translations, travel grants and the publication of journals this dream may not come true.

As a first step, I would recommend that Muslim countries should take an initiative in establishing *institutes of social studies* attached to all or to the major universities. These institutes must be given grants and responsibilities to initiate sociology courses and comparative research programmes.

Today, a significant number of sociologists are working in Western and non-Western countries in various capacities. Their help should be actively sought in developing such programmes. They should be encouraged to join these institutes on a permanent or term basis. It is important that periodical conventions should be held in which these sociologists and other interested persons present their research findings and give recommendations for future developments in teaching as well as in research.

These days, most Muslim countries are paying a great deal of attention to issues of economic and technological development. However, it should be recognized that such issues cannot be resolved in a social vacuum. Moreover, in as much as our main objective is to develop our societies toward prosperity within the framework of Islam, it is crucial that the existing gaps between various social processes and the Islamic ideal should be minimized. In tackling this problem sociologists will play a crucial role provided that the above approach is adopted.

NOTES

1. On Ibn Khaldun, see Mahdi (1957); also for a translation of his *Muqaddimah*, Rosenthal (1958).
2. For an initial statement on this approach, see Parsons (1937 and 1951).
3. Marx's theory is a remarkable synthesis of his ideas derived from the philosophy and the historical studies of his time, the foundation of which can best be found in the *Economic and Philosophical Manuscripts* of 1844. These manuscripts were first published in 1932. For an English translation see Bottomore (1963).
4. See Mead (1934).
5. The Quran describes Islam as Dinal Fitrah which is translated as "the religion of nature". Here I am aware of the way many Muslims interpret this Quranic expression i.e., Islam is in accordance with the nature of man. This interpretation poses a difficult question: what is the nature of man? There is no reason why we cannot interpret Dinal Fitrah as the natural law of human interaction. My interpretation does not merely avoid the above mentioned difficulty. It is consistent with the pursuit and purpose of what is known as science in the most general sense.

BIBLIOGRAPHY

Aron, Raymond, *German Sociology*, (The Free Press of Glenco: Glenco, Illinois) 1964.

Bannister, Gerald P., *Revolution and Displacement: Indonesia After Independence*, (New York: The Social Science Academy) 1973.

Ba-Yunus, Ilyas, *Distribution on Crime and Delinquency in Metropolitan Karachi: An Ecological Analysis*. Pakistan Institute of Urban Studies, Monograph 15: Karachi 1975.

Becker, Howard, "Whose Side are We On", *Social Problems*. Vol. 14, pp. 239-47 1967.

Bottomore, T. B., *Karl Marx: Early Writings*, (London: Watts and Co.) 1963.

Burchard, Waldo W., "Role Conflicts of Military Chaplains", *American Sociologist Review*, 19 (October) 528-35, 1954.

Connor, Walter D., *Deviance in Soviet Society* (New York: Columbia University Press) 1972.

Cuvilliar, A., *Où va la Sociologie Française*, (Paris: Marcel Rivière) 1953.

Direnzo, Gordon J., "Sociology in Italy Today", *International Review of Modern Sociology*. 2 (March) 33-58 1972.

Gouldner, Alvin, *Coming Crisis of Western Sociology*. (New York: Basic Books) 1970.

Hassan, Mohammad, "Rural Urban Differentials in Crime: Egypt under Farouq and Nasser". *International Law Review*. Vol. 6, No. 2 (Fall) 120-27 1977.

Hetcher, M., Review of I. Wallerstein's "The Modern World System", *Contemporary Sociology*. 4 (May) 217–22 1975.

Hill, H., "Peripheral Capitalism: Beyond Dependency and Modernization", *The Australian and New Zealand Journal of Sociology*, 11(1): 30–37 1975.

Inkles, Alex, "Making Man Modern: On the Causes and Consequences of Individual Change in Six Developing Countries", *American Journal of Sociology*, 75 (September) 208–25 1969.

Inkles, Alex and D. Smith, *Becoming Modern: Individual Change in Six Developing countries*. Harvard University Press 1974.

Kamarovsky, Mirra, *Women in the Modern World: Their Education and Their Dilemmas*, (Boston: Little Brown) 1953.

Lynd, Robert S., *Knowledge for What* (Princeton: Princeton University Press) 1939.

Mac Rea, Donald G., *Classics in Sociology*, (New York: Philosophical Library) 1958.

Mannheim, Karl, *Essays in Sociology and Social Psychology*. (London: Routledge and Kegan Paul, Ltd.) 1953.

Marsh, Robert M., *Comparative Sociology*. (New York: Harcourt, Brace and World, Inc.) 1967.

Mead, George Herbert, *Mind, Self and Society* (Chicago: The University of Chicago Press) 1935.

Mahdi, Mohsin, *Ibn Khaldun's Philosophy of History*, (London: George Allen and Unwin, Ltd.) 1957.

Mills, C. Wright, *The Sociological Imagination* (New York: Oxford University Press) 1959.

Moore, Wilbert E., "The Singular and the Plural: The Social Significance of Industrialism Reconsidered", in Nancy Hammon (ed.) *Social Science and the New Societies*, Social Science Research Bureau. Michigan State University, East Lansing 1973.

Moore, Wilbert E., "Modernization and Rationalization: Process and Restraints" in Manning Nash (ed.) *Essays on Economic Development and Culture Change in Honor of Bert F. Hoselitz*, published as a supplement to *Economic Development and Culture Change*, V.25, 1977.

Parsons, Talcott, *The Structure of Social Action*, (Glenco: The Free Press) 1937.
The Social System (Glenco: The Free Press) 1951.

Rosenthal, Franz, *Ibn, Khaldun: The Muqaddimah* (New York, Pantheon Books) 1958.

Turner, Bryan S., *Weber and Islam* (London: Routledge and Kegan Paul) 1974.

Wallerstein, I., *The Modern World System: Capitalist Agriculture and the Origin of World Economy in the Sixteen Century* (New York: Academic Press) 1975.

Chapter Three

An Islamic Concept of History

Abdul Hamid Siddiqui

Abdul Hamid Siddiqui was born in Pakistan in 1923. He held a master's degree in Economics and served as Professor of Economics at Islamic College, Gujranwalah. He was a Fellow of the Islamic Research Academy, Karachi and Associate Editor, *Tarjuman-al-Quran*, Lahore. He died in 1978. His published work includes: *The Life of Muhammad*, Lahore 1969; *Sahih Muslim* (translation and editing), Lahore 1974; *Philosophical Interpretation of History*, Lahore 1969; *Prayers of the Prophet*, Lahore 1968; *Prophethood in Islam*, Lahore 1968.

The Quran is not a book of history but it is a divine verdict on history. The superb style in which the Holy Quran has discussed the different phases of the progress of various nations — their rise, development and decline, as well as the causes underlying these changes — has no parallel in the historical records of the world. It was under the impact of the Quran that man learned to furnish answers to the two fundamental questions "Why did it happen?" and "How did it happen?", that he began to fight against the conception of "chance" as the motive force of change in the universe, and strove to discover the determining law of which, what man calls "chance" is the visible expression. Thus, the transition from mere narration of events to their rational explanation and the introduction of logical order in the recording of them, all these developments in human history are due to the Holy Quran. That the Divine Book narrates the events of the past as *'ibrah* (instructive value) is because of the fact that the events occurring in the past and in the present have meaningful relations between them, or in other words, they have a common law working behind them. Just as there is no change in the laws of nature, and the physical phenomena of our age are controlled by those very laws which governed them in the past, so is the case with the human race and its problems. The passions and pleasures and the political and domestic problems of the people of former ages were, in all likelihood, much the same as ours, as the psychic make-up of all human beings is identical. The Islamic view of history is universal; it is neither time-bound nor space-bound. It does not examine the role of the whole of humanity in its attitude to truth and righteousness.

Further, the Quran and the Sunnah have brought into prominence the role of man as a free agent working within certain limits. The philosophers and historians of our modern age have been stressing unduly the deep, impersonal, unconscious processes that govern social changes. They have accordingly tended to minimize the power of ideas and ideals, or have denied this power altogether. The complexities that make it difficult for them to ascertain the law of social change also strengthen their erroneous impression that man has no real freedom to shape his own history. It is under this false impression that various theories of social change have been evolved and propounded. Philosophers like Spengler look upon culture, which is another name for man-made environment, as an organism governed by the biological laws of life and death. He, in his craze to depersonalize history, seized upon the outer manifestations of culture for building his hypothesis, but ignored the inner, dominating human force working behind it.

Whatever divergence of views there may be between Spengler and Hegel, they base their philosophical systems on a common hypothesis, i.e., man and ideals have no decisive role to play in human history. According to Hegel, the whole drama of life is the unfolding of the world-spirit through a dialectic process. Men may pose a struggle for higher ideals, but all their activities are directed to the goal towards which they are driven by the world-spirit.

The philosphy of history as propounded by Hegel raises the question whether the philosophy conceived on these lines can commend itself to moral reason. He could not provide a satisfactory answer to this question and evaded the issue by simply arguing that the true ethical unit was not the isolated individual but the moral organism, the state or society, in which he was brought up, and that the claims of the latter must take precedence over those of the former. That the individual should completely submerge himself in society and perish for the good of the whole does not strike Hegel as morally outrageous. It is in Hegelian dialectics that all atrocities perpetrated in the course of history found justification since what "exists is alone real", and as such, power as a naked brute force was deified and became the object of adoration, while the individual, "the roof and crown of creation", lost his identity altogether. In other words, the soul of man and its yearnings for immortality and freedom of will became meaningless terms. Human beings were trained not to be morally virtuous but to prove themselves to be efficient cogs in the vast and complex mechanism of society. Man lost all contact with his Creator and thus he deprived himself of the higher values of life.

Marx stripped Hegel's philosophy of its mystical spirit, but retained its dialectical process of social change, and, while substituting the modes of production for the world-spirit, asserted that the ultimate determinant of social change was not to be found in ideas and ideals, but in the changes in those modes. Hence "transitions from one phase of social development to another are not effected because of the demands of new rational principles or of new conceptions of truth and justice, since these belong to the superstructure, and what renders them acceptable is that changes in the production forces create an environment which makes them seem the natural expression of what men have come to desire." Marx's Historical Materialism tends to see the leaders of thought and action merely as the carriers of social forces, ultimately economic. This is how Marxian theory negates all higher strivings in man, reducing him to the ignoble position of a helpless straw before the mighty hurricane of economic forces. It has urged society to suppress brutally individual egotism with the help of collective egotism, so that man may rise in revolt against his moral conscience, against truth and justice and against the Lord of the universe.

Arnold Toynbee has broadened the philosophical basis of history by considering civilizations as the real units of history and not through the conventional approach confined to the career of a society or of a nation. He has tried to emphasize the fact of human creativity, but his frequent appeal to the rhythm of "challenge and response", which makes itself felt over the entire field of action, is, in the words of a philosopher, "nothing but an attempt to cast into the limbo of outworn belief the iron law of fate".

Western philosophers of history, in spite of their striking differences of views, almost agree on the following points:

a) The human ego is hemmed in by time and space and it enjoys no freedom of will.

b) The collective impersonal is alone real and the independent existence of the individual is an illusion.

c) Man's destiny is entirely shaped by social forces and not by his inner being.

d) There is no eternal truth, no objective standard of morality and justice. All these concepts are relative to time and space, and thus there is no law and command that can be held to be universally true.

Islam offers different views on these basic points. Man's physical being is hemmed in by time and space, but not his ego, by virtue of which he has been made the vicegerent of God on earth. He is made the trustee of free personality for which he has been charged with moral responsibilities in regard to all his thoughts and deeds. Morality, essential in Islam, presupposes freedom, ability to choose, select, accept, or repudiate. And where there is no freedom, there can be no morality, for moral conduct applies just to these types of action in respect of which man enjoys the freedom of choice. An action which results from outside pressure has no feeling of morality about it. We appreciate the good deeds of a man, simply because we feel that had he willed otherwise, the opposite course would have been followed.

> And say; It is the truth from your Lord; so let him who will, believe; and let him who will, disbelieve. (18.29)

> Lo! We created man from a drop of thickened fluid to test him; so We make him hearing, knowing. Lo! We have shown him the way whether he be grateful or disbelieving. (76: 2, 3)

> Man shall have nothing but what he strives for. (53:39)

> Allah does not change the condition of a people until they change their inner selves. (8:54, 13:11)

According to the Quran, man's activity is neither the unfolding of the world-spirit nor is it determined by the modes of production in which he lives, but by his own ego.

Again the Quran and the Sunnah reject the view that the collective impersonal is alone real and the separate existence of the individual is an illusion. Everything is real according to the Quran and there is a divine purpose behind its creation; and man is exhorted to bear this fact in mind, that "everything that exists has a serious purpose of its own and it has not been created in vain". (3:191)

> And We created not the heavens and the earth and all that is between them in sport. We created them not save with a (Divine) purpose, but most of them know not. (44:38, 39)

Man has been created as a social being in the sense that it is in society that his piety and God-consciousness can achieve their full possibilities.

That he exists merely for the material benefits of society, or that his life has a meaning only when he becomes a mere cog in the vast machine of society, is wrong. Were we to believe that the individual exists only for society, we should have to infer that the significant and important thing in life is not the spiritual development of souls, but the social development of communities. This thesis, on which the modern social set-up of so many nations is based, debases man and robs him of the dignity with which God has invested him. The treatment of the individual as a will-less automaton in the stern hands of the community is a denial of the personal relation between him and God, a substitution of the worship of God by the worship of the community.

The Islamic view of relationship between man and God is that man stands in the presence of God as an individual and in this very capacity he will have to explain his conduct on the Day of Judgement.

> *It will be said unto him*: "Read thy book. Sufficient is thy own soul as the reckoner against thee this Day. He who follows the right way follows it for the good of his own soul; and he who goes astray does so only to his own loss. And no bearer of burden shall bear the burden of another." (17:14, 15)
>
> "And now you come to Us one by one, as We created you at first." (6:94)

Man's relationships with God is primarily in his individual capacity, but Islam stresses the point that this relationship should externalize itself in a just social order, free of all forms of tyranny and oppression. The religious piety of man, no doubt, represents his individual attitude, but this attitude must find its expression in a healthy social set-up wherein no injustice is done to anyone. It means that salvation which a Muslim longs for can be attained when he paves the way for the salvation of his fellow-beings, nay for the entire human race. The standards of values and the tasks which he, in his religious devotion, learns to recognize as an inescapable necessity are found to be duties not for him alone but for the whole of mankind. This fact has been beautifully explained in a saying of the Holy Prophet (peace and blessings of Allah be upon him):

It is reported on the authority of Hadrat 'Aisha (Allah be pleased with her) that Allah's Messenger (peace and blessings of Allah be upon him) once observed: "An army would attack Ka'bah. But when it would be at *Baida*, its vanguard and rear would sink in the earth." Upon this, Hadrat 'Aisha (Allah be pleased with her) said: "Allah's Messenger, how will it be that (the whole army including) its vanguard and rear would be made to sink in the earth, whereas there would be some among them who would be forced against their own will (to join the army)?" The Holy Prophet (peace and blessings of Allah be upon him) said: "The vanguard and the rear would be made to sink, and they would be made to rise according to their intention on the Day of Resurrection (agreed upon)."

Thus Allah will treat human beings as individuals in the Hereafter, but as members of society in this mortal world, He will reward them for their good deeds and punish them for their misdeeds, taking into consideration their individual behaviour in the life beyond the grave, but will confer favour on them or inflict chastisement upon them collectively in this material world.

Hence the self-consciousness of the Muslim bears an active and social quality as well. This means that whatever a Muslim does in life has its value in history, and its worth can be judged from the type of manhood it has created and the good it has brought to the maximum number of people. That is the reason why so much stress has been laid on reflecting over the past and present experience of mankind and on seeing for ourselves as individuals and as the members of the world-*Ummah* of Islam, the moral and social outlook that led to the rise of mankind as well as the ethical disintegration which brought about its ruin.

> "Already before your time have precedents been made. Traverse the earth then, and see what has been the end of those who denied the signs of God. (3:136)"

> "See they not how many generations We destroyed before them whom We had established upon the earth in strength, such as We have not given to you? For whom We poured out rain from the skies in abundance, and gave streams flowing beneath their feet? Yet for their sins We destroyed them; and raised in their wake fresh generations (to succeed)". (6:6)

> "And how many generations We destroyed before them who had better possessions and appearances? Say, 'For him who is in error, the Beneficent will extend the rope to them until they see the warning of God being fulfilled either in punishment or in the approach of the Hour. Then they will know who is worse in position and weaker in force." (19:74–75)

Islam exhorts a Muslim to see not only the outward manifestation of the different happenings of human life, but to study the undercurrent of ideals and motives which have shaped those happenings. The historical references and the accounts of the past are given in the Quran not so much to fill in the gaps of our factual knowledge, but to systematize and generalize it and to take lessons from it. The Holy Quran treats of the events of the past not only with a view to reviving them in our memory but to making them meaningful and instructive to us. It selects the significant events, interprets them in the light of moral law, and then evaluates them according to ethical judgements; and in the whole process of selecting, interpreting and evaluating the facts, it provides answers to the crucial questions about the destiny of mankind.

The attitude of Islam towards historical knowledge is of great significance in human understanding. The Muslim historians showed great exactitude in surveying the entire course of human development, but they sought to determine its origin and goal as well. This can be illustrated from the fact that the famous world-history of Ibn Khaldun is entitled *Kitab-ul-'Ibr*. The *'Ibr* (moral lessons) here stands as a prominent key word which reveals the underlying idea for which history was studied by the Muslims.

The Holy Quran and the Hadith have urged us to review past events, both reported and experienced, for the purposes of awakening in us a strong moral sense and enhancing our ability to act according to the Commands of God. The purpose is to penetrate into the otherwise meaningless succession of events and discern the ever-present design and will of the Creator and perceive that all life and events are the outcome of His conscious, all-embracing Power. Unless one is in spiritual accord with the demands of that Power, man cannot fulfil the Divine purpose for which he has been sent to this world.

This feeling and perception have in fact awakened in man an urge for permanent moral values in life, for without the existence of such absolute standards of moral evaluation, he cannot judge for himself whether his aims and actions are in conformity with the demands of the planning Will or not. In the absence of such eternal standards the concept of morality loses all its precision and becomes the handmaid of expediency, to be interpreted arbitrarily according to one's personal or communal needs and the changing requirements of time and economic environment.

The Holy Quran states clearly that just as the laws of nature are immutable and universal, so are the laws of religion. They are the objective realities of life, independent of ephemeral changes in man's environment.

> He (Allah) has prescribed for you the religion which He enjoined on Noah and which We have now revealed to thee, and which We enjoined on Abraham and Moses and Jesus. (42:13)

Thus to a Muslim, his *din* is not a mere passing phase in his history, but the ultimate source of ethics and morality, law and politics, economics and metaphysics—the very alpha and omega of all his yearnings and aspirations in every age and under all sets of conditions.

According to the Quran, God does not unfold Himself in history as Western peoples believe. There is no denying the fact that it is not without a purpose that He gives dominance to certain people at one time and deprives them of this position at another. This ebb and rise of the fortune or misfortune of the people has a Divine purpose to serve. The Quran observes:

> If a wound has afflicted you, a wound like it has also afflicted the disbelieving people. We bring these days to men by turn, that Allah may know those who believe and take witnesses from among you and Allah loves not the wrongdoers. (3:140)

The use of historical events as a warning against certain patterns of action, and exhortation to adopt a certain course of life, is not an absolutely new concept introduced by Islam. All the revealed books have clear indications of this but the way which the Muslim historians observed objectivity and exactitude in the narration of events, and deepened their moral aspects, is a purely post-Islamic development. The works of Ibn Khaldun furnish an enviable example of this moral-scientific approach to history.

Islam has also given to the world the fundamental principles of historical criticism. Accuracy in recording facts which constitute the materials of history is an indispensable condition of history. Sincere and accurate knowledge of facts ultimately depends on those who report them. Thus the very first principle of historical criticism is that the reporter's personal character is an important factor in judging his testimony.

> O believers! If a bad man comes to you with a report, verify it. (49:6)

It was from the application of the principle embodied in this verse to the reports of the Prophet's traditions that the canons of historical criticism gradually evolved. As for the moral lessons to be drawn from the events and used as warnings for the future, the contributions of the Muslims are unrivalled.

The growth of the historical sense in Islam is a fascinating subject. It was because of this sense that the Muslim scholars tried to observe the signs of God not only in nature but also in history, and made it an avenue to the understanding of the ultimate reality.

Chapter Four
On Islamizing the Discipline of Psychology

Abdul Hamid al-Hashimi

Abdul Hamid al-Hashimi was born in Damascus in 1922; Syrian; Chairman of the Dept. of Psychology, Faculty of Education, King Abdulaziz University; Formerly Cultural delegate and member of the King Abdulaziz University Council; Ph.D. in Psychology, University of Karachi, Pakistan; publications include *"Formative Psychology: Fundamentals and Applications from Birth to Old Age"* and *Individual Differences: Analytical, Applied Study in the Fields of Education and Sociology."* Also published five articles on Education and Reading, Hay-ibn-Yaqthaan, Freud and The Effects of Anaesthetics; member of the American Psychological Association in Washington; participated in the First Conference on Teacher Training and contributed in the Council of the Educational Psychology Research Centre. He was in charge of the Secretariat of the Council of Higher Studies at the Faculty of Education; speaks Arabic, English and French.

I. *The Discipline and Its Early Development*

Psychology discusses claims that probably represent the earliest scientific attempts of man on this earth to understand himself and the different changes that he experiences: when he feels joyful, sad, contented, jealous or altruistic; when he remembers or forgets, likes others or dislikes them, and when he tries to discover the reality of other humans in their anger or joy, forgetfulness or recollection.

Despite its ancient origins, however, psychology is a relatively modern science — modern in its methodology, especially in its employment of laboratory means which are more objective, adaptable, accurate and standard, particularly in this last quarter of the twentieth century.[1]

During its history of emergence and development psychology has changed and modified its spheres of interest as a result of the efforts of a long series of ancient and modern psychologists. At present this science is endeavouring to crystallize its broad domain, demarcate its frontiers, and increase its information about the human self. It asserts at the same time, and with scientific frankness, that it is just beginning to comprehend human behaviour in its different dimensions and that a great deal remains for it to observe, discover, study, test and apply. Bearing this in mind, the present study will pursue the following broad, clear lines:

a) A general survey of the development of the field and frontiers of psychology with a view to finding out whether psychologists have exhausted their fields of enquiry, or whether they still have new horizons that call for more investigation and realization, and a review of the chief problems of modern psychology.

b) An inquiry into Islam's attitude to psychological studies; whether it is passive and disapproving, neutral, or positive and encouraging.

Contemporary psychology, as it exists in both East and West, is founded upon a fundamental principle accepted as almost axiomatic. This principle is that man in his psychological structure is absolutely dominated by his physical, organic and material structure. Accordingly, the behaviour of the human being is dominated by a physical equation whose essence is a material stimulus and a physical response. Man formulated his life in a certain way to serve his material ends, namely, his physical well-being and material prosperity. This prevalent psychological trend reveals itself in different degrees, ways, and terms in the works of most psychologists of this century as well as of the last. Two separate movements have dominated the scene: the Russian concept in the East, and the American concept in the West.

The Russian concept is represented by Ivan Petrovitch Pavlov (1849–1928), a physician who specialized in the physiology of digestion, and Vladimir Bekhterev (1857–1927), a neurologist. They claimed that the so-called "associated" or "conditioned reflex action" dominates the psychological life of the human being, and they disregarded all other factors such as consciousness, the unconscious, drives and the will. According to this organic biological concept, the psychological life of man is based on the brain with its two functions:

a) A sensory function which consists of partial stimulants resulting from physical movements to which the human being is constantly prone. The brain is thus like a receiving set.

b) A dynamic function which consists in controlling the conditioned reflexive movements. This conditioning is located in the cerebrospinal cortex.

This purely physical concept of the human psyche overlooks all other factors such as feeling, drives, individual or species inheritance. To say the least, this concept strips man of one of the principal characteristics which distinguish him from other living beings: desire, free will, free choice and a sense of responsibility. Psycho-biological inevitability is dominant in animal psychology, but it is restricted in human psychological life. The error of such a purely organic concept of the human psyche lies in giving one explanation to the various aspects of psychological life.[2] The Russian physiological concept was subsequently affected by the philosophy of Karl Marx—dialectic materialism—when the latter introduced economic, material factors along with the biological in the study of the human psyche, as has been clearly shown by K. N. Kornilor (born 1879), director of the Institute of Applied Psychology in Moscow District University.[3]

Another materialistic, natural concept characterized the Western trend in general and the American trend in particular. Indeed, it defined psychology as a natural science that deals with visible kinetic behaviour. Assuming psychological life to consist of behaviour, and the latter to be summed up in the formula Stimulus—Response, or S—R, it called itself the "science of behaviour". This concept is best represented by John Broads Watson (1878–1958) who denied the existence of consciousness, the subconscious, or the spontaneous drives which are deeply hidden in the human psyche. In the beginning, this concept attracted the attention of psychologists by some of its limited achievements, but faced difficulties when it tried to explain sensation, memory, recollection and thought. The worthlessness of this "visualization" can be seen, for instance, in its explanation of memory, which it describes as being partly the general responses on the part of the individual, and partly a feeling of joy or suffering which emanates from sensory stimuli in genital organs and other erotic regions of the body. The explanation of thought, according to this concept, is even more absurd, for it considers it an internal speech in which the vocal chords and other normal speech organs vibrate. This purely materialistic trend has been refuted by many psychologists. John Dewey (1859–1952), pioneer of the school of Functional Psychology, criticized the idea of the reflex, and emphasized the psychological faculties (capacities), refusing to reduce the latter to stimulus and response.[4]

Finally, the concept of sexual repression emerged as a pivot for humanistic psychological studies in Europe as they were carried out by Sigmund Freud (1856–1939) towards the end of the last century and early in the twentieth. Freud founded psychoanalysis on the theory of sexual repression and other pertinent ideas, chief among which are: sexual energy, the unconscious, transference, the Oedipus complex, the theory of Libido and the Electra complex. All the ideas under this concept might be summed up in three words: repressed infantile sexuality. According to this concept, psychological life is centred upon the principle of pleasure, whether as means or end. The concept was an immediate success. However, a while later, grave scientific criticisms were levelled against it.

One of these criticisms is that this concept formulates theories, then tries to find for them a basis in experience through analysis, and finally illustrates those theories by examples drawn from patient histories. In addition it tries to account for psychological life by a partial and one-sided explanation, that is, the sexual drive. In fact, recourse to this concept reflects the peculiarities and abnormalities of European society, especially its widespread sexual promiscuity. It also reflects the values of European Jews. Many contemporary psychologists have undertaken to refute this concept, the most scholarly of them being Alfred Adler (1870–1937).[5]

II. *The Chronic Problem of Psychological Studies*

The historical development of psychology briefly outlined above was an essential factor in the emergence of the diverse concepts described above. More than a thousand years ago, psychological studies began with speculations about the soul. Centuries later, psychology moved from inquiry into the nature of the soul to a study of the psyche and the reality behind it. Thence, it moved to mental processes. Later still, it emphasized feelings and senses. Finally, it shifted to the study of consciousness and the unconscious. Today its attention is almost entirely given to behaviour.

The development of psychology and the shift in its centre of interest are best illustrated as follows:
1. The first stage: *The soul*
2. The second stage: *The psyche*

3. The third stage: *The mind*
4. The fourth stage: *Feeling and Sensation*
5. The fifth stage: *The unconscious and the unconscious mind*
6. The sixth stage: *Visible behaviour.*

Describing the development of these concepts, Woodworth said: "Psychology first lost its 'soul', then its 'mind', then its 'feelings', then its 'consciousness', then its 'unconsciousness' and only its 'visible behaviour' is now reserved for it." It is possible that this development is merely a reflection of the dominant cultural background of Western psychologists. At first, Christian culture prevailed, then it was weakened because the Church, which monopolized it, resisted scientific progress. Later on, the industrial revolution and materialistic culture had a far-reaching impact; and there was also the influence of European secularism. The emergence of these concepts, however, harmonizes with the recent detachment of psychology from philosophy, as well as with the Western theory of man, the sole subject-matter of psychology. It should be no wonder that the horizons of psychological studies have not yet been clearly defined.

The basic problem of modern Western psychology is that it does not regard man as a whole in all his constituent dimensions and psychic phenomena. Rather, each stage in the history of psychology has singled out one aspect of the human psyche and devoted complete attention to it, ignoring other aspects. Such bias was the result of the diversity of cultures and orientation of psychologists. Its consequence was the appearance of numerous dissimilar designs and concepts, all of which cannot be true; nor can any single one be true. The problem with psychological studies has been, and still is, their failure to comprehend accurately, profoundly and thoroughly the various aspects of the human psyche, and to maintain the balance among all interacting factors in psychological life.

III. *The Effects of Defective Perception of Psychic Reality*

The following are some of the deviations which defective perception of psychic reality such as we have found in Western psychology brings in its trail:

a) The confinement of studies to one aspect, differing from one psychologist to another, gave rise to generally confused or distorted pictures of psychology, in spite of some success in several practical fields.

b) Unreserved emphasis on the intellectual or mental dimension led to the distorted view of man as a calculator or computer. The view achieved intellectual and scientific progress; but it was unbalanced.

c) Too much interest in the physical aspects of the psyche, in its drives and activity, led to disregard of the spiritual and moral aspects which compose a major part of personality. This shortcoming resulted in anxiety along with material prosperity, in emotional loss along with intellectual advancement. That is why contemporary man in the West is a disfigured giant who has the power to give great impetus to scientific and material development, but suffers from anxiety, aimlessness and lack of self-awareness. He knows a great deal about the depths of the oceans, the mountains of the moon and the valleys of Mars, but he still stands at the threshold of any real knowledge of his own psyche, the nearest thing to him in the whole universe. His material needs seem to have deflected him from seeking to understand himself. "Man", Alexis Carrel wrote, "has progressed in the domain of material interests. He has also acquainted himself with some secrets about the structure of matter, and in this way has been able to have control over many things, barring himself. Man is still unknown ... and our knowledge about ourselves is still primitive and partial".[6]

IV. *The Rise of Proper Psychological Trends*

From among numerous and conflicting psychological studies and concepts, new trends began to develop which sought to introduce some scientific balance between the various constituent elements of the human psyche.

Among these trends, we note:

a) If the history of psychology is divided into an old stage, a middle stage, and a modern stage, Descartes (1596–1650) can be said to have founded the modern stage. He envisaged the human body as a machine dominated by the psyche and steered by the soul, and defined psychic life as the outcome of these aspects without underrating any of them.

b) The principle of insight has become accepted as a psychological fact in certain types of learning, and as a feature of the human soul lying beyond the limits of perception and observation.

c) The principle of purposivism (hormic psychology) has been established and declares that man's behaviour is purposeful, that purpose is not necessarily materialistic or biological, and that drives are not all materialistic, physical or biological.

d) The principle of range has appeared, enabling psychology to regard man as composed not of a single structural aspect, but of a complexus of interacting factors which are not independent of each other, thus underlining the dynamic nature of psychic life.

e) New horizons and themes in psychological studies have been discovered — creative thinking, self inspiration, psychology or morals (character), child morality, etc.

It should be added here that much of what psychologists have contributed may not be scientifically erroneous in itself. Rather, their shortcomings lie in their assumption that a single concept is the ultimate and only truth about the human self, or in proceeding from one discovery to generalize about the human psyche as a whole. Although contemporary trends in psychology and recent studies have shown that the existing concept of psychology is imperfect and not thorough, and have pointed to the need for further exploration, the general scientific idea about the human psyche is still partial and unbalanced. We still need a general scientific development in the field of psychology.

V. *Psychology in the Muslim World*

We have so far summarized the condition of psychological studies in the twentieth century, an era in which the West has borne the torch of civilization, while Muslims were awakening from their slumber. After the First World War, Muslims did open their eyes, though only intermittently, to find themselves invaded and ruled by an alien power determined to subject them militarily, economically, socially, scientifically and intellectually. As a result of defeat, Muslims reacted by imitating the victorious invader.

The West had indeed planned to alienate the Muslim people from their Arabic and Islamic identity. The Muslims sent their students to learn in the West. Studying in the West became the hope of every intellectual, for the acquisition of knowledge in any branch of learning. Those who were not lucky enough to study there kept dreaming of the West, consoling themselves by reading Western books in translation.

In psychology imitation was the rule in that period. Our young men returned from the West and proudly repeated what they had learned, without distinguishing the worthwhile from the worthless. In many cases, they translated the books they had studied and used them as textbooks, unaware that the psychological situations and cases they were teaching to their students were French, English, American and Russian, not Muslim.

The divisions which separated the various strands of the Western tradition reappeared among Muslim psychologists. Graduates of German schools emphasized Gestalt psychology; those of English schools stressed drives, the psycho-hormical purpose, and mental processes. American-trained psychologists favoured analytical and functional psychology. Graduates of the French schools were more interested in the areas of consciousness, intelligence and personality, while those of the Russian schools insisted on biological and physiological elements, stressing the principle of correlative conditioning and economic and social factors in the formulation of personality.

It should be remarked here that most psychology departments and laboratories in Western universities, particularly the American ones, are dominated by Jewish scholars committed to their own peculiar Jewish existence and culture. Most Arabic books on psychology which have appeared in the last forty years reveal that this period has been one of uncritical assimilation and copying. The majority of that generation of Muslim psychologists were ignorant of the Islamic legacy, thus complicating the problem further. The chronic problems of psychological studies in the West were transplanted in us with all the West's hostile and scornful disregard for the influence of the spiritual and moral in the formation of the human psyche in general and of the Muslim psyche in particular.

VI. *Current Problems of Psychological Studies*

Besides the foregoing shortcomings, other problems surfaced in the course of the development of psychology in the Muslim World, namely: those pertaining to psychology, methodology and instruments; those pertaining to the scope of psychological research; and those pertaining to the objectives of psychology.

A. *Methodological Problems of Psychology*

As a result of foreign training, some of our psychologists applied Western psychological theories and principles in their field studies in Islamic societies. In a recent enquiry about the methodological problems confronting Muslim psychologists, Dr. Louis Kamel Matekal pointed out that the problem was one of using psychological experiments designed for foreign societies. In the special field he investigated, that of measuring intelligence, he noted that Muslim psychologists passively conceded Western psychologists' stereotyped description of non-Westerners' face-to-face thinking, lack of abstraction, tendency to limitation and lack of originality. They were not aware that the culture factors used in the measurement vitiated the results; that complete reliance on and excessive use of the statistical method in psychological measurement led inadvertently to treating the human psyche as a material object with fixed qualities, or as a computing machine. The fact that Western psychological statisticians had exposed such methodological shortcomings had escaped them.[7] Moreover, they overlooked the differences of personality features arising from differences on the socio-cultural level.

It is therefore our responsibility to develop the means of data collection and the standardization of psychological tests in a way that is suited to our society as a whole or to a certain portion thereof, and with reference to our contemporary needs and expectations.[8]

B. *Problems Pertaining to Limitation of the Field*

Western psychological studies usually confine their field of interest to a single aspect of man. As mentioned above, psychological studies at various times in the past focused on the soul and nothing else, then conventional psychology focused on the mind only. Some psychological schools emphasized a single aspect, like the physical drives, the sexual instinct, the apparent behaviour (motor-behaviour), the correlative conditioning reflex; and this partiality is true of the various trends of so-called modern psychology. This narrow conception of the human psyche, with its unbalanced emphasis on one or another psychological constituent, produced studies and findings lacking scientific accuracy which mirrored the West's shortsighted view of man.[9]

C. *Problems pertaining to Narrowness of Objectives*

It took psychology a very long time to crystallize its objectives. It began as a theoretical and contemplative activity aiming at satisfying man's curiosity and desire to understand himself. At that time, it was a branch of philosophy. When information concerning the human psyche increased, psychologists tried to predict the behaviour of individuals.[10] Later on they began to apply their findings in the fields of management, industry, commerce, leadership, jurisdiction and criminology. They also entered the field of war and politics.

VII. *Islam's Concern for Study of the Human Psyche*

Firstly, Islam holds that the study of the human psyche is a useful scientific activity, worthy of encouragement like all knowledge. There can be no doubt that the Quran is not a book of psychological, scientific, geographical or cosmic theory. It is essentially a book of guidance, disclosing enough of the secrets of the psyche and the universe to teach that man may seek to understand and use the universe in direct obedience to Allah.[11] The Quran discloses the potentialities and imperfections of the human psyche, and contrasts its states of guidance and misguidance. Hence, the word *nafs* (self, psyche, etc.) occurs more than 300 times in the Quran in dozens of contexts.[12]

The study of the human self is apparently the shortest and surest way to faith in Allah, the Creator of man, Who proclaimed: "Soon will We show them our Signs in the farthest regions of the earth, and in their own souls, until it becomes manifest to them that this is the truth . . . as also in your own selves; will ye not then see?" (Qur'an, 41:53; 51:21).

Secondly, Islam has taken care to dissect the human psyche into its various constituent aspects — its emotions, consciousness, attitudes, functions and numerous faculties — in order to facilitate the education of man, for whom we are responsible, both his childhood and his adulthood. Since human education is enjoined upon us, the methods which secure its success are enjoined as well. Every useful branch of knowledge is imperative in Islam. Its pursuit is either a personal or a collective task, but it is always an intrinsic religious duty. It is the personal responsibility of every Muslim to acquire such knowledge as is needed for his personal purity of faith and perfection of worship. On the other hand, those branches of knowledge which are a prerequisite for the welfare of society as a whole — the arts and sciences, industry, medicine, engineering, agriculture, chemistry, physics, biology, psychology, etc. — are society's responsibility.

Thirdly, many eminent scholars carried out significant psychological studies. They approached man as an integral whole: his body, soul, consciousness, emotion and behaviour — a concept which has recently become the target of modern psychological studies.[13]

VIII. *Islam's View of the Scope of Psychology*

The subject matter of psychology is the psychological structure of the human being in both the physical and the spiritual realms. Just as the physical aspect of man has its drives, potentialities, stages of development, repression and consummation, restfulness and tiredness, well-being and illness, the spiritual aspect also has its demands and needs. It is influenced by factors that bring about its well-being or suffering, and make the individual happy or unhappy. In addition to the physical and spiritual aspects, the Quranic notion expressed in the word "*Sawwaytuhu*" (15:29; 38:72) indicates three psychological structural dimensions:

a) Comprehensiveness: The human being has in his structure both the physical and the spiritual aspects simultaneously and constantly. He is not a mere animal that only breathes, moves and grows. He has also a spirit sustained by belief in Allah, and finds his peace and happiness in having contact with God. Nor is the human spirit one that merely soars in space; for man is and remains both body and soul. The soul by itself is not man, nor is the body by itself man.

b) The notion "*Sawwaytuhu*" also refers to the organic balance among the constituent aspects of the soul. The physical aspect has its importance, limitations and potentialities, so it should not be neglected in favour of desires and drives. At the same time the latter should not be indulged in to excess, or at the expense of other dimensions. Similarly, the emotional aspect must not be neglected or suppressed because it is the expression of life, but it should not be given preference over the human psyche, or over the mental aspect, for instance. The same applies to the spiritual aspect which should not be disregarded or dismissed from the psychological life of man, but must not be pursued to the detriment of other constituents. Man is a human being and not an angel. The realization of relative balance in the psychic life is of utmost importance for good health in the different stages of man's life. In fact, most psychological disorders result from the loss or lack of that relative balance among the constituent aspects of the human being: body and soul, mind and emotion, social and moral aspects, desire and spirit. When the human being behaves in a certain way, he does not do so as a robot, but as a result of spiritual, cultural, mental and physical factors.

IX. *Psychology and Morality*

Man's neglect of his psychological guidance and righteousness has brought about a moral crisis of severe gravity. Moralists turn to the psychologists in their attempt to diagnose this great problem of modern man. The modern psychologist, however, has nothing to offer. Indeed, he persists in his position that morality does not concern him. In Europe and America, psychology has decided to stay aloof from an area which should have been the main part of its study and research. Most moralists have no confidence in psychologists. They in fact fear the influence of psychology in the field of values and morality. In his attempt to understand the motives behind moral values, the psychologist has often destroyed those very values, even though he may not have been deliberately working for their destruction.[14] It is regrettable that in its present state, psychology can provide no help to the moralist. It is still a crude science whose results are under examination, a science which has gone too far in its study of psychic diseases, with the result that it judges the integrated, balanced, and adjusted man by using the same scale that is used for the abnormal and maladjusted. In this way it has distorted the general outlook of the human psyche.

Moreover, some psychological researches have been manipulated by military and imperialistic institutions which have used their findings in the fields of war — whether hot war, cold war or economic war. Such manipulation has led to the establishment of the following as physiological principles: "divide and rule", "Starve out your dog, and it follows you", and "The weapons of intelligence are three: Sex, money and wine". This has led many unbiased moralists to describe psychology as a two-edged weapon. The honest, straightforward person would use it for a virtuous purpose, and the abnormal, sick person would manipulate it for his own base, sinister ends.

As was pointed out earlier, psychology has passed through numerous stages of growth. Today, when a science called applied psychology has emerged in which we speak of psycho-therapy, psycho-pathology, industrial, administrative, military, judicial, criminal and social psychology, it is absolutely necessary that psychology should no longer escape its duty and that it should resolve to tackle in its applied form man's values, morals and virtues.

As psychology is interested in attitudes, mental processes, and the visible behaviour of the individual, it should not neglect values and virtues as stimulants, movers and ends for man. Inanimate objects cannot be described in ethical terms; but no analysis of human behaviour is complete without such a description. The Muslim psychologist is bound to assume moral judgements in his psychic analyses. For his purpose, man's emotions are either proper, balanced, adequate, or excessive, abnormal and misplaced; constructive and positive, or perverse. Behaviour is, in his eyes, always either good or bad. In the sphere of the development of conscience, of will and habit, the mental make-up, i.e. values as ends and objectives, is a crucial constituent of self-consciousness, of the process enabling the individual to exercise control, to strive and aspire, and to modify his outlook accordingly. This demands the emergence of a science called "moral psychology", "psychology of ethics", "psychology of values", or "the moral developments of the psyche", because moral values are the facts of the emotional and mental life of all humans. Our psychological studies will never be complete unless they enter the sphere of values and morals, provided that these values be spiritual, God-inspired and in accordance with man's innate nature or *fitrah*. There are in the psyche inborn (or innate) inclinations adaptable for goodness and truth. It is only through careful and patient care, from the moment of birth until death, that those inclinations can be cultivated and developed. The Almighty said: "By the soul, and the proportion given to it; and the enlightenment as to its wrong and its right" (91:7–8). His proclamation, "Truly he succeeds that purifies it [the soul] and he truly fails that corrupts it" (91:9–10), obviously refers to the process of conscious and purposive control aiming at enhancing and reinforcing, at conducting and steering, behaviour.

X. *General Framework for the Field of Psychology*

We may now proceed to draw up a general plan for the study of the different constituent aspects of the human psyche, beginning with those with which the modern schools agree, and then turning to those which, though controversial, are nonetheless necessary for the integrity of the human psyche.

A. *The Psycho-Sensory Field*

The study of the parts of the human body which are directly related to the emotional and mental processes — like the senses, the nervous system and the glands — and of the stages of physical growth and their distinctive characterization.

B. *Intellectual Dimensions*

The child's intellectual characteristics; stages of mental development at various chronological and mental ages. Intelligence, ability, and personal differences. Learning, its fundamental and supplementary process; thinking; imagery and imagination; recollection, reminiscence, and forgetfulness. The problems of intellectual dimensions and their solutions.

C. *Emotional Dimensions*

Emotions: their types, causes, and manifestations, love, anger, fear and jealousy; the drives. Emotional development at various stages of life according to chronological age. Emotional disorders. Requirements of a healthy emotional life.

D. *The Social Dimensions*

Psycho-social tendencies: friendship, leadership, blind-following, cooperation, competition and contention. From I (the individual) to WE (the group). Development of psycho-social behaviour and life-stages; requirements and problems.

E. *The Spiritual Dimensions*

Faith and its positive psychological effect on the integrity of personality. Belief in God as a psychological instinct; man–God relationship. Stages of development of religious consciousness at different ages. Worship and its behavioural, social and emotional effects. Atheism and spiritual loss: case studies.

F. *The Moral Dimensions*

Diverse moral tendencies and their importance in the formation of habits and feelings. Stages of development of moral consciousness as related to chronological and social growth. Effects of an ideology based on faith in God and moral flawlessness. Factors of moral development, of control and guidance. Moral deviation; materialistic and pragmatic morality; selfishness. The study of cases and their treatment.

At the end of this general framework for psychology it should be pointed out that when a detailed plan of any branch of psychological application is prepared, the following two points should be borne in mind: throughout the foregoing curriculum, care should be taken not to return to philosophical duality as in the Platonic concept of the human psyche. That duality was based on the isolation of the soul from the body and the materialistic concept of the soul. The Islamic concept is based on the interaction and interdependence of body and soul, of mind and emotion. Our psychological study of the spiritual dimensions does not imply a return to philosophy and its theories about the soul. It is rather an applied and analytic study which would result in practical clues for the psychological well-being of the individual and an understanding of the factors that secure that well-being.

XI. *The Objectives of Islamic Psychology*

In our psychological studies and their applied form we should bear firmly in mind that behaviour, function, and objective are not synonymous with each other. Behaviour is the external and internal expressive action during the emotional situation or the cognitive process; function is the direct provisional result realized through the behaviour of the various organs of the body; and objective is the desired aim which steers the psychological drives towards that behaviour, is partially fulfilled during the psychological activity, and is fully fulfilled after it. Moreover, some objectives are instrumental in achieving others.

The objectives of Islamic psychology may be described as follows:

(a) Understanding present behaviour and its motives as they are. To this end, every useful method of investigation has to be applied: observation, introspection, experimentation and measurement.

(b) Intelligent speculation about future behaviour, in order to predict deviance or aberration before it occurs and demands treatment.

(c) Spontaneous control of drives, phenomena and purposes. Being intelligent and responsible, man can and ought to, mobilize and direct his motives. He should not escape responsibility for the goodness or badness of his behaviour, high or low level of his objectives. The difference is indeed great between the process of repression which denies the existence of the drives or tries to conceal them, thus causing diseases and disorders, and the process of "rational government" and "control" which directs natural energy to attain satisfaction and quiescence.

(d) Directing all the drives, emotions, and mental operations to win God's favour, i.e., to realize that noble, supreme stage of human existence characterized by continuous harmony with God's patterns. The satisfaction of natural drives is good and does not go against piety so long as it follows the ways allowed in Islam. Thus, eating, drinking, sex, possession, joy, sadness, contentment, anger, learning, recollection, and recreation are so many acts of piety and worship offered to Almighty God if they do not violate the teachings of Islam and are undertaken with the intention of generating the energy necessary to do good deeds. They become the basis for the conventional forms of worship, for prayer, fasting, charity, etc. In this way the emotional, mental and behavioural life of the human psyche becomes, according to Islam, one long, continuous act of worship. God Almighty says: "But teach (thy Message): for teaching benefits the believers. I have only created Jinn and men, that they may serve Me' (51:55–56).

It might be objected that to consider spiritual and moral development as a subject of psychological study goes against the nature of psychology which is concerned merely with matters and aspects inherently and directly pertinent to the psychological make-up of man. This objection might arise from the assumption that the inherent fields which bear a direct relation to the psychological composition of man are his inherited aspects. This is a false assumption which hardly any modern psychologist accepts. It might arise out of a reluctance to extend scientific research to areas declared to lie beyond investigation and analysis. This is a dogmatic position, contradicted by an equally noble and precise set of data, namely, the spiritual. Man in his very *fitrah* is a living being who contemplates and hopes, who believes and has faith. He perceives the certainty of the existence and oneness of God. He perceives the importance of being obedient to God. This spiritual aspect should be studied by psychology in order to establish its nature, the stages of its development compared with the chronological age of the human being, the factors securing its well-being, and the treatment of weakness of deviation. From his first day on earth as a genus, and from his first years as an individual, the human being has had a preference for truth and candour in his personal relations. He is pleased with sincerity and he has a tendency to feel ashamed when he commits something wrong. These inborn human tendencies are central to the field of psychology, and this science should study their origin, the stages of their normal development, the factors that lead to their deviation or weakness, and the means to assist their treatment and promotion.

It might be countered, in response to the foregoing argument, that to regard the moral and spiritual aspects as psychological dimensions in the same sense as the emotional and mental aspects, would bring psychology back to the domain of philosophy, because the subject of the soul is central to the latter, and psychology has for centuries endeavoured to free itself from philosophy and its conflicting theories. On the face of it, the objection seems scientific and reasonable. However, on close inspection it demonstrates its spuriousness. Just as the physicist's ignorance of the essence and nature of electricity, magnetism and the atom does not prevent him from studying their effects as heat, motion and light, so the psychologist who studies intelligence, emotion and joy defining them as "power", "energy", "charge", "force", "ability", or "inclination" is not hampered by his ignorance of their nature and essence. He studies intelligence and emotion by studying their manifestations, factors, and effects. It is the same with the spiritual aspect of a man's psychological make-up. Psychologists need not study the reality or essence of value and spirit, but the manifestations of them in the life of man, man's fulfilment or otherwise of their axiological content.

Contemporary psychologists agree that the major drawback of traditional psychology, when it was still a branch of philosophy, was that it confined its attention to the spirit and did not consider perception, feeling, emotion, sensation, etc. Likewise, it is an obvious scientific shortcoming of modern psychology that it ignores the spiritual dimension of man, despite the latter's tremendous influence upon human life.

XII. *Islamic Psychology and the University*

Voices are being heard throughout the Muslim World that psychological studies should be undertaken in the light of Islam. Some psycho-researchers have already begun work in a few institutes and universities. A proposal, once presented by Dr. Fuad al Ahwani in his preface to 'Abdul Karim 'Uthman's *Psychological Studies by Classical Muslim Scholars*, envisaged the establishment of a new discipline in the educational curriculum, to be called "Islamic Psychology".[15] In most Muslim universities, some branches of psychology have come under scientific and medical specialization. However, psychology is still an Arts subject despite its attempts to be an objective science by subjecting its data to repeated testing and to measurement expressed in numbers. Its passage to the "Science" division could be achieved without objection if it could be established that such a move was not designed to relieve it of its duty to consider the moral and spiritual aspects as proper material for study, falling well within its jurisdiction.

XIII. *Recommendations*

The following recommendations are presented for possible adoption and implementation:

(a) Muslim psychologists should be asked to devise genuine psychological tests to reveal and test the various manifestations of psychological activity of the Muslim individual. These tests should be completely original, not translations of foreign tests with a slight modification to give them a different appearance.

(b) Islamic psychological centres should be established in capitals and main cities all over the Muslim World to carry out research on various age groups and in various psychological activities. These centres should have psychological laboratories to test individuals and standardize tests.

(c) Conferences and seminars should be held in which committed Muslim psychology researchers exchange skills and ideas, prepare plans for new psychological researches, and circulate them to various centres and institutions.

(d) An international Islamic psychological journal in Arabic and English should be established to publish studies made by Muslim psychologists. Muslim students should be encouraged in various fields of psychology at university level. Students who hold a B.A. degree in any field should be admitted, after taking certain basic courses, for M.A. and Ph.D. degrees in psychology.

(e) Short courses (say, of three months), should be offered to contemporary psychologists and teachers of psychology at various levels with the purpose of providing them with a grounding in Islamic spiritual education, along with a detailed and scientific explanation of the Islamic concept of the human psyche.

(f) All Muslim universities and colleges should be asked to establish departments of psychology, so that psychology can be studied in the light of Islamic studies, and that students specializing in Islamic sciences can become familiar with psychological studies.

(g) An international committee for Muslim psychologists should be set up with headquarters and a permanent secretariat in Mecca. The committee should undertake to publish the journal recommended above, and it should hold conferences and seminars, organize training courses, and provide financial support for research, and for the publication and distribution of psychological works produced in accordance with the moral and spiritual imperatives of Islam.

NOTES

1. Morris Roclain, *A History of Psychology*, translated by Ali Zai'our and Ali Mukalled, published by 'Uwaidat, Beirut, first edition 1972, p. 9.
2. G. L. Flogel, *Psychology in a Hundred Years*, translated by Lufti Fatim, first edition, Darut-Tali'ah, Beirut, 1973, p. 175.
3. Robert Woodworth, *Modern Schools of Psychology*, translated by Kamal Dusuki, pp. 119–122, 320–21, 161–62, Pub. Darul-Ma'arif, Egypt, 1948, First Edition.
4. Ali Zai'our, *Schools of Contemporary Psychology, Darul-Andalus*, Beirut, First Edition, 1971, p. 289. Also see *Encyclopaedia*, "Behaviourism", Vol. 3, p. 398.
5. For more details see "Freud under Trial" by the present writer in the *Journal of King Abdulaziz University*, First year, First issue, (1395 A.H., 1975) pp. 381–405.
6. Alexis Carrel, a scientist specializing in chemistry, biology and physiology in his book *Man, the Unknown*, which is an important and extensive work in this subject.
7. Dr. Sayyid Muhammad Khayri, *Statistics in Psychological Educational and Sociological Studies*, 4th Edition, 1970, Darul-Nahda al-Hadithah, Essay beginning p. 37. Also see Dr. Fuad al-Bahiy al-Sayyid, *Statistical Psychology*, Darul Fikr al-Arabi, 1971, p. 636.
8. For more details see Dr. Louis Kamal Malikah, "*Readings in Social Psychology in Arab States*,' Vol. 2, 1970, El-Hay'ah ah al-Misriyah al-Armmah Lit-Ta'lif Wan-Nashr (General Egyptian Commission for writing and publication), pp. 17–18 and more details in ch. 4 pp. 62–82.
9. Dr. Mustafa Mahmud, *Some Secrets in the Qur'an*, Kutubul-Yawm, No. 115, Sep. 1976, pp. 60–67.
10. Deopold B. Van Delin, *Methodology in Education and Psychology*, translated by Muhammad Nabil Nawfal, Sulayman al-Khudari al-Shaykh, and Tal'at Mansur Gabrial, Anglo-Egyptian Bookstore, 1398 A.H. 1969 A.C. pp. 71–7.
11. Muhammed Qutb, *Studies about the Human Psyche*, 1387 A.H. (1967 A.C.) pp. 8–9.
12. The word *"an-nafs"* (which is most often used to render the meaning of psyche or self) occurs 305 times in the Qur'an in more than forty surahs.

(1) Meaning Allah, the Divine Entity, the term occurs in the following verses:
"Say Peace be on you: your Lord hath inscribed for *Himself* the rule of mercy: verily if any of you did evil in ignorance, and thereafter repented and amended his conduct, lo! He is oft-forgiving, most Merciful" (6:54). "But God cautions you to remember *Himself*" (2:30).

(2) Meaning the human entity, the human being as an integral whole, the term occurs in the following verses:
"On no *Soul* doth God place a burden greater than it can bear. It gets every good that it earns, and it suffers every ill that it earns" (2:286). "Do you enjoin right conduct on the people, and forget to practise it *yourselves*" (2:44). In one other verse, the term occurs twice, once in the first sense and once in the second. Reporting on God's questioning of Jesus, the Quran says: 'He will say: Glory to Thee! Never could I say what I had no right to say. Had I said such a thing, Thou wouldst indeed have known it. Thou knowest what is in *my heart*, though I know not what is in *Thine*. For Thou knowest in full all that is hidden" (5:119).

(3) Meaning Adam, the father of mankind, the term occurs in the following verses:
"O mankind! reverence your Guardian-Lord, who created you from a single *Person*, created of like manner his mate, and from them twain, scattered like seeds countless men and women, reverence God, through Whom you demand your mutual rights, and reverence the wombs that bore you; for God ever watches over you" (4:1).
"It is He Who hath produced you from a single *person*: Here is a place of sojourn and a place of departure" (6:98).

(4) Meaning man's conscience, intention and internal talk, the term occurs in the following verses: "To God belongeth all that is in the heavens and on earth. Whether you show what is in *your minds* or conceal it, God called you to account for it" (2:284). "And they say to *themselves*, Why does not God punish us for our words?" (58:8).

(5) Meaning the evil-prone psyche, the dark side of the self when it induces the individual to evil, sin, and to deviant, abnormal, mean or whimsical behaviour, as in the story of Yusuf (peace be on him) and the overlord of Egypt and his wife, after Yusuf's innocence was unanimously acknowledged: "This say I, in order that he [Pharaoh] may know that I have never been false to him in his absence, and that God will never guide the snare of the false ones. Nor do I absolve my own self of blame, the *human soul* is certainly prone to evil, unless my Lord do bestow His mercy: but surely my Lord is oft-forgiving, most merciful" (12:52–53). "They follow nothing but conjecture and what *their souls* desire! Even though there has already come to them guidance from their Lord!" (53:23).

(6) Meaning that self-conscious psyche which reproaches the individual for wrong-doing or shortcomings (the kind of psyche which wakes after sleep), the term occurs in the following verses: "I do call to witness the Resurrection Day; and I do call to witness the self-reproaching soul! Eschew evil! Does man think that We cannot assemble his bones? Nay, We are able to put together in perfect order the very tips of his fingers" (75:1–4).

(7) Meaning the contented psyche which is blessed, with whom God is pleased and which is pleased with Him, as in the verses, "To the righteous soul will be said, O thou *soul, in complete rest and satisfaction*, come back thou to thy Lord—well pleased thyself and well pleasing unto Him" (89:27–28).

(8) In other passages, the term refers to those aspects of personality which develop after birth like patterns of behaviour, thinking processes, and moral systems, as in the following verse: "Because God will never change the grace which He hath bestowed on a people until they change what is in *their own souls*: And verily God is He Who heareth and knoweth all things" (13:11).

(9) Meaning the individual's drives, needs, and desires, whether innate and inherited though subject to reinforcement and modification, or acquired under the subject's personal responsibility, the term occurs in the following verses: "By the *soul*, and the proportion and order given to it, and the enlightenment as to its wrong and its right;—truly he succeeds that purifies it, and he fails that corrupts it!" (91:7–10). "And for such as had entertained the fear of standing before their Lord's tribunal and has restrained *their soul* from lower desires, their abode will be the Garden" (79:40–41).

(10) Meaning the soul, the essence of biological and physical life, the loss of which is tantamount to the loss of life, the term occurs in the following verse: "It is God that takes the *souls* of men at death; and those that die not He takes during their sleep. Those on whom He has passed the decree of death, He keeps back from returning to life, but the rest He sends to their bodies for a term appointed. Verily in this are signs for those who reflect" (39:42).

(11) Referring to the physical and organic qualities which differentiate males from females, the term is applied to human beings, to plants and even to some inanimate being, such as the negative and positive in the fields of electricity, the atom, etc. It occurs in the following verse: "Glory to God, Who created in pairs all things, whether out of the earth or out of their own selves (souls), and other things of which they have no knowledge" (36:36).

    13. See Muhammad Qutb, *Studies about the Human Psyche*, 1387 A.H./1967 A.C., an excellent reference on the subject, particularly pp. 13–70.
Notable among Muslim phychologists of the classical period were the following:
Abu Hamid Muhammad ibn Muhammad al Ghazali (d. 504 A.H./1111 A.C.), known

as Hujjat al Islam (Advocate of Islam) in his book *Ihya' 'Ulum al Din* (The Revivification of the Sciences of Religion), particularly in the Third Volume, "The Redemptive Deeds". In this book, al Ghazali spoke about the psyche, the soldiers of the heart, the diseases of the heart, their symptoms, and cures. He also wrote about habit, its various stages, and its types. He designated psychological studies, "the science of Treatment."

Abu al Faraj 'Abd al Rahman Ibn al Jawzi (d. 597 A.H./1201 A.C.), in his book *The Intelligent*, which is composed of 33 chapters. In this book the writer discusses the differences between each of the following terms and between these terms and others; the brain, comprehension, and intelligence. He also discusses measuring intelligence by means of certain physical and behavioural signs. Then he quotes instances and cases among children, men, women, the wise, the insane, even animals.

Abu Nasr al Farabi (d. 329 A.H./950 A.C.) in his book *Opinions of the Citizens of Utopia*. He discusses the human psyche, man's need for being with others, the importance of leadership, and the traits of the leader.

Ibn Sina (d. 432 A.H./1037 A.C.), a physician on whom were conferred the honourable titles "Al Shaykh" and "Al Ra'is", in his books *The Moods of the Psyche* and *Hints and Notices*. He discussed the psychological forces, the hidden and visible senses. He carried out a pioneering study into consciousness; its types and factors, and the relationship between the psyche and the body. He also touched on adult psychology.

Abu 'Abdullah Ibn al Qayyim al Jawziyyah (d. 751 A.H.) in his book *The Rescue of the Anxious Man from the Snares of the Devil*. In the first chapters he classified the hearts of men as healthy, sick or dead, and described the various ways of heart-treatment, including the treatment of depression. He also discussed the relationship between the body and the psyche, and their reciprocal influence when they suffer pain or distress.

Abu Hayyan al Tawhidi (d. 403 A.H.) in his book *Al Muqabasat*, a portion of which he devoted to "Friends and Friendship". In this book he tackled awakening, sleep and dreams, perception, anger, and the faculty of distinguishing different colours, especially black and white. He also discussed laughter and its causes, and thoroughly investigated the truth about friendship and the friend.

Abu Bakr Muhammad al Qays Ibn Tufayal (d. 581 A.H./1185 A.C.), author of *Hayy ibn Yaqzan*, an extensive study of the major psychological stages in Hayy's life: as a child, a youth, a young man, and an adult. Here the author discussed the various psychological phenomena of Hayy — imitation, love, imagination and recollection, and other activities such as discovery and experimentation within himself, in his surroundings and in the universe, his reaction to disasters, his behaviour as a learner (with an explanation of the best way for adults to learn a language), and his longing to meet with his fellow human beings. See also by the writer of this article the extensive discussion of the same topic under the title "Hayy ibn Yakzan, a Pioneering Psychological Study and an Islamic Masterpiece," *Journal of the Faculty of Education*, Makkah, the first issue, Shawwal, 1395 A.H./October, 1975 A.C.

14. See John Karl Flogel, psychologist (1884–1955), who was at home in five languages both written and spoken, *Man, Morality and Society* One Thousand Books (No. 628), translated by Othman Nawiyyah and Dr. Sa'ad al Ghazali, Dar al Fikr al 'Arabi, Cairo, 1966, pp. 15–17.

15. Abd al Karim 'Uthman, *Psychological Studies by Classical Muslim Scholars*, first edition 1382 A.H./1963 A.C., Preface by Dr. Ahwani, pp. 3–4.

16. The psychological studies mentioned here are offered by the Department of Psychology, Makkah, for postgraduate students.

Chapter Five
Restructuring the Study of Economics in Muslim Universities

Mohammad Nejatullah Siddiqui

Mohammad Nejatullah Siddiqui was born 1931, Indian; Professor, Department of Islamic Studies, Aligarh Muslim University, Aligarh—India. Professor, King Abdulaziz University, Jedda. Ph.D. in Economics, and Arabic Education at Rampur and Madrassatul Islah, Azamgarh. Editor, Quarterly *Islamic Thought*, Aligarh (1955–59); Assistant Director, Islamic Research Circle, Aligarh. Publications include Urdu Translations of Abu Yusuf: *Kitab al Khiraj* and Syed Qutb: *al' Adalah al Ijtima 'iyah fi'l Islam*. Author of *Recent Theories of Profit* (Asia—Bombay); *Banking Without Interest* (Lahore); *Economic Enterprise in Islam* (Lahore, Delhi); *Some Aspects of Islamic Economy* (Delhi, Lahore). *Principles of Partnership and Profit-Sharing in Islam* (in Urdu: Delhi, Lahore) *Islam's Theory of Property* (2 volumes in Urdu: Lahore) *Insurance in Islamic Economy* (in Urdu: Delhi), *The Road to Islamic Renaissance* (in Urdu: Delhi), *Muslim Personal Law—Proceedings of a Seminar* (edited—Delhi) and *Bibliography of Islamic Economics* (Islamic Foundation, U.K.)

A critique of the present state of teaching of economics at University level in Muslim countries should focus mainly on two points: how far it is relevant to the actual working of our economies and, how far it is in harmony with our aspiration to transform these into Islamic economies. Such a critique would be the proper starting point in suggesting changes that would harness the teaching of economics for proper management of these economies and for accelerating their development along Islamic lines. It may be said that economics is primarily an understanding of how the economy functions, that the business of transformation and development can be taken up only at a later stage. The implication is that the latter should not be allowed to interfere with the former. Two points can be made to meet this objection. Firstly, it must be our economies that we understand and explain, not some other economy or even a hypothetical one. Secondly, what we try to understand and analyse has the potentiality to change. As Muslims do have clear preferences on the desired direction of change, this potentiality should also be explored with the desired changes in view. The teaching of economics in Muslim countries should, therefore, be reviewed and its content and style changed to make it serve the cause of Islamic transformation and the economic development of these countries.

Modern economics which we are presently teaching is in fact capitalist economics which is largely based on the British and, more recently, on the American experience. As J. R. Hicks said on receiving the Nobel Prize for Economics in November 1972, "Economics is to so large an extent a British Science." The point has also been made earlier by Joan Robinson who, while introducing her *Exercises in Economic Analysis*, argued that "English Economists, from Ricardo to Keynes, have been accustomed to assume as a tacitly accepted background the institutions and problems of the England each of his own day; when their works are studied in other climes and other periods by readers who import other assumptions, a great deal of confusion and argument at crosspurposes arises in consequence".

Standard economics is supposed to explain the *modus operandi* of an idealized capitalist economy. Its concepts of man, property, freedom, competition, and the role it envisages for the state, are all derived from the particular ethos and cultural milieu of Eighteenth and Nineteenth century England. Far from being universal, some of these concepts are quite irrelevant for present-day Muslim societies. Following its basic postulate that pursuit of self-interest by each individual redounds to the benefit of all, it focuses its attention on the material and technical problem of maximum production and ignores the human and moral problem of just distribution. Thus modern capitalist economics has three main faults: it treats particular attitudes and institutions as if they were universal, which they are not; it suppresses the moral problem and extols the technical; and, lastly, it almost ignores the powerful agency of the state and its positive role in economic management and development.

It is a fact that economics evolved in an era of colonization and imperialism. If the economics of the closed society suffers from the defect of idealizing the emergent British merchant and industrialist, suppressing the problem of distribution and immobilizing the state, the application of economic theory to international trade and international finance is biased in favour of the colonizers, the bankers and financiers and the imperial powers. It is not only inadequate to handle the problems of the developing countries of the Third World, its value for them is largely negative. Despite the claims of this form of economics to be universal in its application, a number of eminent economists who have been concerned with the problems of development, notably Lewis, Schultz, Kuznets, Galbraith and Myrdal, have emphasized its inadequacies from the view-point of the Third World.

There is no need, however, to dwell too long on the inadequacies of modern economics in a general way. Let us turn to some of its basic concepts and see how they are value-loaded, oriented to individualism, exploitative capitalism and imperialism, and strongly biased against a social and moral approach to economic affairs. We shall select a few of them: economic rationality, capital, loan and development.

Before we examine these concepts it would be useful to expose the fallacy that a suitable dose of socialist economics will redress the wrong done by exclusive reliance on modern economics. Marx did not visualise a man different from the one his predecessors had in view. Socialist economics is distinguished by its elevation of the state to the position of the chief economic agent which owns all resources, allocates them and manages production and distribution in a planned manner. The market is replaced by the state as the mechanism for choice, though partially. Man, the individual, becomes a passive agent not to be trusted with freedom. This swing from one extreme to another is neither realistic nor does it suit the temper of the Islamic peoples. This approach is not very relevant to present-day Muslim economics which is far from being socialistic. It is not in harmony with the desired Islamic direction of change either.

It might be argued that socialist economics gives priority to distribution over production. But it is not a moral priority rooted in human attitudes. It is a mechanical priority exclusively relying on coercion.

The basic concepts of Marxist economics, namely labour, surplus value and classes, are concepts destructive of the capitalist system; but they are poor ingredients for a constructive programme of action. A critical examination of these concepts would reveal their ideological bias and convince us that they do not fit within the conceptual framework of Islamic economics. They are of little use in understanding our own economic problems. Limitations of time and space do not permit us to go into greater detail so far as a critical review of socialist concepts is concerned. We shall therefore proceed to examine the capitalist concepts mentioned above.

Some Alien Concepts

Erratic and inconsistent behaviour is not amenable to scientific analysis leading to generalizations that could have some predictive value for groups but not for individuals. This is possible only with consistent behaviour resulting from deliberate choice guided by certain norms. All intelligent persons try to be rational in this sense, their success depending on the availability of relevant information. It is reasonable, therefore, to assume rational behaviour as the basis of economic analysis of households and firms. But *economic rationality* assumed by modern economics goes farther than that. It regards maximization of advantage for the self as the chief attribute of all economic agents. The consumer maximizes utility or satisfaction and the firm maximizes profits. The concepts are strongly individualistic and utilitarian. Altruism or one caring for the good of others in one's choice is ruled out. The time horizon is narrowed down to the near future if not the present. A care for life in the hereafter is thus excluded, however genuine one's faith in it. The concept tends to regard non-economic ends irrelevant for choice, in so far as utility, satisfaction and profits are interpreted in purely economic terms. As maximization of profits implies minimization of costs, the focus is on direct cost to the individual, ignoring ecological and socio-cultural costs. Thus the concept of economic rationality enshrines the materialistic individualistic wordly approach to life in vogue in the hey-day of capitalism. Those were the days of aversion to religion and morality and revolt against the Church. The Darwinian idea of survival of the fittest further sanctified exploitation of the weak by the strong in the name of economic rationality. The idea of superiority of the white race justified imperialism.

The concept of economic rationality is not suitable for the analysis of behaviour in Muslim countries, for two main reasons. It is not realistic. Men do care for the good of others besides caring for their own good. Their choice is influenced by their assessment of its possible consequences for society, for their neighbours, etc. They do care for the after life. Their choice as consumers and their decision as producers, employers, traders etc. is influenced by their faith or lack of faith in the life hereafter. They are also motivated by social, cultural and political considerations. These tendencies may be strong or weak, depending on the extent of available information and the strength of the relevant commitment. Yet the tendencies are there. A concept that ignores these tendencies can hardly do justice to reality. Once the concept of *economic rationality* is given the status of a basic assumption it has a tendency to serve as a norm. Men tend to regard it as the proper way to behave. That is the difference between human sciences and natural sciences, the object in the latter remaining unaffected by any assumption the scientist makes regarding its behaviour. No wonder that altruism, other-worldly considerations and non-economic motives tend to be regarded as irrational modes of behaviour by those who take economics seriously. When economic theory is applied to practical problems and policy prescriptions are made, this bias results in the neglect of the vital interests of society and its individuals. This is the second reason why Western "economic rationality" cannot be admitted into the conceptual framework of Islamic economics. It would hinder the process of Islamic transformation by virtue of its normative role. Islamic economists have suggested its replacement by the concept of "Islamic rationality" implying orientation of action towards maximal conformity with the Islamic norms.[1] This concept too is normative in nature, but it would serve as a better tool of analysis if due allowance is made for weakness of the Islamic commitment in present day Muslim societies. It is a broader concept than that of economic rationality and a closer approximation to the actual reality. Its normative role will promote the cause of Islamic transformation of our societies.

Capital

Another crucial concept of modern economics that does not suit the conceptual framework of Islamic economics is that of capital as a separate factor of production independent of enterprise. Islam does not recognize capital's claim to a guaranteed positive return in the form of rate of interest. It does however, recognize its claim to a share in the profits of the enterprise in which it is invested provided it is exposed to the risks of the enterprise and assumes its share of the loss, if any. Capital borrowed with guarantee of repayment, irrespective of the results of the enterprise, has no claim to a share in its profits. If factors of production are defined in the context of their claim to a share in the net product, risk capital is a factor of production and loan capital is not.

That, of course, is the position regarding money capital. Money capital ensures command over capital goods such as machinery, tools and buildings, etc. Modern economics regards the latter as productive and then, in view of the capacity of money capital to command capital goods, attributes productivity to it also. In doing so it slurs over the crucial difference between risk capital and loan capital. This is not legitimate as the two enter the process of production in entirely different ways.

Whether the use of capital goods in production will produce a value larger than their own value depends on the success or failure of enterprise in a world characterized by uncertainty of market values. The uncertain world in which capital and enterprise co-operate in production does not guarantee a positive value productivity to either of them. The practice of treating interest as an item of cost and ascribing it to the productivity of capital is based on the particular institutional set-up of the capitalist society. All attempts to prove value productivity of capital goods have foundered because capital can be measured only in terms of value and that requires prior knowledge of the rate of interest which the marginal value productivity is supposed to *determine*. In fact it is the institution of interest that makes it possible for capitalist economists to ascribe a positive return to capital, that is its value productivity, and not the other way round. The concept of capital as a factor of production independent of enterprise makes this fallacy plausible. This fallacy must be exposed. This concept is a tool of exploitation and not a tool of scientific analysis. As a tool of analysis its role is to legitimize the institution of interest. It confuses our understanding of the economic process in which the value outcome of capital goods is uncertain, by ascribing a positive value productivity to it.

While discarding this concept we do not propose to shut our eyes to the obvious fact that physical capital enters the process of production in a manner different from that of labour or enterprise. It has a distinct identity and a distinct role. These, however, are technical facts. Physical capital is a distinct factor of production in the technical sense. The phenomenon of diminishing marginal physical productivity of *capital goods* is also a technical truth. The downward slope of the demand curve for a particular good, e.g. a machine, reflects this truth. Its upward sloping supply curve reflects rising costs in the short run. The market price of a machine is determined by supply and demand. It is an item of cost for the enterprise in which it is employed. The fact that it remains employed over a period of time is, *economically*, relevant for the money capital involved in procuring it and for the machine itself.

When we consider the money capital tied up in the process of production it is of no consequence what part of it is used to procure labour or other ingredients. What is crucial in case of money capital invested in enterprise is whether it is exposed to its risks or not. Scientific analysis requires the distinction between risk capital and loan capital to take this difference into account.

Loan

The concept of loan has undergone a significant change under capitalism. In the pre-capitalist era loans represented real savings. Command over social produce acquired by making some contribution to it was temporarily transferred to the borrower. As the propensity to consume of the borrower is higher than that of the lender, the act of lending had an antideflationary effect which contributed to the health of the economy.

With the advent of banking a loan became so much purchasing power created and passed on to the borrower. It constitutes an additional claim to the stock of goods and services available in the society. It is not matched by any prior or simultaneous contribution to the social product. It is therefore inflationary in its impact. In so far as bank loans are for consumption no addition to the social product is expected even in the future. In the case of productive loans an increase in production may take place after a time lag provided idle resources exist in the economy. A continuous stream of bank loans is bound to exert an inflationary trend in view of this time lag. A society which lacks idle resources or faces serious bottlenecks in the supply of capital goods, skill or raw materials required will be the worst sufferer in this regard.

If, in the pre-capitalist era, the borrower had to pay an additional sum of money as "interest" besides repaying the principle, it involved a simple transfer of command over real resources from one class of people to another. Payment of interest on money created *ad hoc* for lending magnifies this effect to such a great extent that its very nature changes. If the reserve ratio is 10 per cent so that saving deposits worth 100 enable the bank to advance 1000 as loans, and the annual rate of interest is also 10 per cent, the amount transferred will be 100 per year. Lending of the 100 saved, without intervention of banks and creation of money by them, would, however, involve a transfer of 10 per year only. While the repayment of the bank loan will extinguish the money the bank has created, the interest paid to the bank will remain. The redistributive role of the institution of interest on loans has therefore become much too decisive with the advent of banking.

Notwithstanding the fallacy of the various justifications offered for the saver's claim to interest, the new situation calls for a fresh justification for the lender's claim to interest when the money lent is created in the process of lending, exists, as long as the loan exists, and is extinguished as soon as the loan is repaid.

It is a social convention that enables this remarkable thing to happen, i.e. the habit of people to keep their cash with the banks and deal in cheques, their confidence in the banking system being ultimately rooted in the protection given by the central bank. It is society which permits new purchasing power (unaccompanied by a simultaneous addition to social production), to be created and tolerates the inflationary pressure involved in the hope that it will lead to a subsequent increase in the social product through utilization of idle resources.

Why does the economics of capitalism fail to emphasize the obvious difference between loans that represent real savings and loans that do not? Why does it treat the creation of additional claims to social production on a par with a mere transfer of existing claims? Why is the same concept supposed to cover two dissimilar entities? The reason lies in the legitimization of interest on bank loans which the cover of the old concept provides. Once the social origin of the privilege extended to borrowers is recognized society is bound to claim the resulting benefits, whatever they are. The banks could not be entitled to more than service charges, with an extra amount to cover the risk of non-repayment if it is not taken care of in a different manner. What the banks presently appropriate is much more than that.

It does not suit the conceptual framework of Islamic economics to treat the two kinds of loan alike. Lending proper must be distinguished from the social permission for exercising additional purchasing power in anticipation of additional production. From the Islamic point the act of lending is an act of charity, a good deed. It is not an act of business motivated by profits. The borrower is not obliged to pay back anything over and above the principal borrowed. Lending belongs to the area of altruistic cooperation, an important dimension of economic activity in Islam. The social permission for exercising additional command over resources in anticipation of additional production is a new phenomenon. The rights and duties of the parties involved have not been defined by the text of Islamic law; this has to be done now in accordance with the spirit of that law. It is evident, however, that they would be different from those laid down for "loans" in the earlier sense.

Development

We need not dwell too long on the narrowness of the capitalist concept of development as much has been recently written on the subject. It was conceived in terms of higher production to the neglect of distribution. One must ask whose development was being talked about. Was it the development of man or that of matter? How can the development of human society be conceived of in terms of additions to wealth only, irrespective of whether these are available to the bulk of its members or not. A rising GNP and increasing poverty, both in absolute and relative terms, can well, and indeed do, go together.

It is not only distributive justice that the capitalist concept of development ignores. It also ignores the ecological and social costs of development such as pollution, depletion of non-renewable resources, and stresses and strains on the individual and the family. It fails to take into consideration man's relationship with nature and the quality of his cultural life. The narrow concept of development held sway for more than a century and it is only now that *some* economists are seriously attending to the other inalienable dimensions of development. The reason for this failure lies in individualism which formed the basis of economic thinking. A comprehensive concept of development requires a broader frame of reference including all members of society, future generations, other living creatures and the non-economic interests of mankind. It is only such a broad concept that can suit Islamic economics.

If we must question such concepts as capital and development, what about money, income, employment, banking, insurance and a host of other concepts which are closely linked to the concepts exposed above and are the products of the same materialistic individualistic approach? The need for a critical review of all these concepts is established and this task must have priority with Islamic economists.

Structure of a Course: Some Guide-lines

Shall we, therefore, abandon teaching modern economic theories and their applications in such important areas as development planning, public finance, controls, monetary policy, international trade, etc? This, of course, is neither possible nor desirable. What is needed is a judicious selection of the more enduring elements in the corpus of modern economics and the handling of them in a critical manner. This should be coupled with the injection of Islamic goals and preferences in the applied areas and a reference to the actual economic conditions of our own economies at relevant points. As we do so new concepts will evolve and the old ones will be suitably modified to suit our needs and aspirations. But the successful culmination of this process requires patience and perseverence, besides imagination and commitment to our ideals. Ideological groups have a tendency to neglect somewhat the technical aspect of the matter in the first flush of their exuberance. We should be mature enough not to commit this mistake. The structure of courses outlined below guards against this possibility while giving due attention to the need for reform and innovation. Economics is presently taught in the universities of the Muslim world mostly at the undergraduate level. But many of these universities are planning to start graduate courses in economics in the near future. The time is therefore opportune to focus attention on the structure of courses at graduate level. The University of Cairo and universities in other Muslim countries with a long experience of graduate teaching and research in economics could also review their syllabi and restructure their courses in the light of what follows.

Of great importance is the need to integrate the teaching of Islamic economics with the teaching of economics in general. Nothing short of a fusion of Islamic elements with the valid and enduring elements of modern economics will answer our needs. This is what is suggested in the scheme outlined below.

Price theory has always been the core of any course in economic analysis and it should continue to be so. Laws of supply and demand should be taught with a sceptical attitude towards the various explanations offered to provide them with a psychological basis. Production, cost and revenue functions, elasticities, equilibrium price, simple and discriminating monopoly, price output trends in imperfect markets, etc. should be given a treatment that takes into consideration alternative assumptions regarding the ends of households and firms, paying special attention to the ethically informed behaviour pattern. The concept of derived demand should be taught but the application of price mechanism to income categories such as wages, rent, interest and profits should be modified so as to give due weight to the moral, social and institutional factors involved in distribution. Earlier, we have underlined the need for a review of capital as a separate factor of production. How this and the other factors of production are rewarded in the capitalist economy is of some practical interest. How they are actually rewarded in our own economies is, however, a different question. Still more important is the enquiry into how they would be rewarded in the modified attitudinal and institutional framework of an Islamic economy.

Macroeconomics, comprising the theories of money, income and employment, is another core subject in modern economics. The basic Keynesian tools of analysis need to be handled with a clear comprehension of the type of economy to which this box of tools applies. Employment and output, effective demand, consumer's function, multiplier, savings investment relationship, accelerator and the nature of economic fluctuations, etc. are concepts whose interpretation and relevance is bound to be affected in a context different from that in which they have evolved. In the analysis of aggregate demand and aggregate supply as determinants of employment and output due place has to be given to such social forces as the popular will to develop and state initiative and leadership. The role of money and banking in a *zakat* based economy should also figure in macroeconomic analysis. This is an example of the fusion of Islamic elements with basic economic analysis that is required.

Keynesian macroeconomics, like all the other theories of capitalist economics, is production oriented and neglects distribution. Macroeconomic theories of distribution hardly rectify this imbalance as they operate within the same basic framework. Elements of institutional economics can be profitably introduced in this analysis. The impact of attitudinal changes prompted by Islamic norms of behaviour should provide another area of enquiry in this regard.

Microeconomics and macroeconomics have their application in a variety of areas such as welfare economics, growth, development and planning, finance, trade and international economic relations. But the element of relativity is larger in these branches. Their content would have to be gradually replaced by what is more relevant to developing countries in general and Muslim countries in particular. The received doctrines should be treated as doctrines of capitalism; and the possible alternatives should be spelled out with a view to facilitating the choice of what answers the need of our own economies. The tools of analysis and techniques of economic management should be tested on the criteria of their relevance for developing economies and harmony with Islamic goals.

There is a strong need for a comprehensive course in public economics, which would replace the traditional course in public finance. Such a course should deal with fiscal policy, debt management, physical controls and other instruments of policy through which the state can manage the economy with a view to securing social ends. The economics of public undertakings should also figure in this course. Special attention should be paid to the lessons that can be learned from the socialist experience of managing production, consumption and exchange. Public economics is a subject on which eminent Islamic thinkers such as Ibn Taymiyah (1261–1328 A.D.) have pronounced themselves and due attention needs to be paid to their views. The fast expanding public sector in Muslim countries further underlines the need for such a course.

A course on growth and developmental planning is also necessary. Various models of growth and possible strategies of development should be identified and their social, cultural and institutional implications explained. As regards the techniques of planning, emphasis should be placed on those suited to the home economy with an awareness of the alternatives operative elsewhere. The essential characteristics of an Islamic strategy of economic development should be spelled out with a critical review of the models suggested so far.

A course on comparative economic systems will bring home to the student the point that there exist alternative ways of economic management and the possibilities are by no means exhausted by the models currently operative. A variety of approaches within the systems of capitalism, socialism and mixed economies should be highlighted. Salient features of the economic system of Islam should be presented in such a manner that the possibility of variety in interpretation is not missed by the student.

The study of economic history should focus on past human experience in different regions, special attention being paid to British, American, Russian and Japanese history. An analytical approach to the economic history of diverse peoples at different times broadens the vision and prepares the ground for effecting a synthesis and for launching a new experiment as opposed to feeling committed to a particular model as something inevitable and inescapable. In our universities, the economic history of the Islamic peoples, especially those in West Asia and North Africa, deserves special attention, in addition to the economic history of the home country. The subject could be covered in a series out of which each student would have to select a few.

The history of economic thought has been omitted from the core courses in most universities. It is now taught as an optional subject. We could adopt the same practice. Classical political economy, marginalism, neo-classicism and Keynesian economics, and Neo-Marxian as well as some of the other "heterodox" schools of thought, should be studied in a series of optional courses. This group of courses should include one on the evolution of Muslim economic thinking with special emphasis on the contributions of Abu Yusuf, Ibn Taymiyah, Ibn Khaldun and Shah Waliyyullah. This course should focus on the interaction between Islamic injunctions and changing economic conditions through the ages. This subject has suffered neglect by the historians of economic thought but is one which encourages research and teaching in Islamic universities.

An important subject for introduction at undergraduate level and to be pursued in greater detail at graduate level is the structure of the economy of the home country, its evolution and its current problems. This in fact should be the hub around which all other courses should revolve. It should be one of the compulsory courses.

Area studies, especially those concerning the countries of the Muslim world, are necessary in view of the need for regional planning and greater coordination between the economies of any single region. A series of optional courses should cover the neighbouring countries and the other economies of the region.

A comprehensive course on international economics comprising international trade, multinationals, world monetary system and financial institutions, economics of aid and foreign investment is one of vital importance for Muslim countries. The subject should be treated from the viewpoint of the Third World and due attention should be paid to the implications of the New Economic Order advocated by it. This too should be a compulsory subject.

There is no controversy surrounding the application of quantitative methods in economics which is taught in mathematical economics, statistics and econometrics. But these courses should be designed to answer the requirements of the analysis and management of our own economies rather than to attain formal excellence. They should be aids to clearer formulation of concepts and better comprehension of such interrelationships as are amenable to quantification, with the awareness that much of crucial importance escapes this net. In view of the growing use of quantitative methods in economics, an elementary course on statistical methods and quantitative techniques should be included within the core syllabus, the rest being covered in a series of advanced courses to be offered as optional subjects. As a prerequisite to the compulsory course at graduate level there should be one at undergraduate level in mathematics covering, besides simple algebra, differential calculus and analytic geometry.

The list of optional courses should further include various courses in applied economics, economics of particular industries, economics of technology and innovation, financial organization of society, comparative banking systems, labour economics, population studies, agricultural economics, transport economics and so on. This list can be extended or abridged, depending on the actual needs of each country. It is only in this manner that the present courses gradually become adapted to our real needs and aspirations. Some of them evolve into entirely new ones. This process has, however, to be hastened by the introduction of two more courses as core subjects, one on methodology and the other on Islamic economics.

The course of methodology should expose the methodology of neo-classical Keynesian modern economics, as well as that of the Marxian and Neo-Marxian economics. This should be followed by a review of some important attempts to devise a new methodology. Lastly, a possible Islamic methodology should be discussed in the light of progress made so far by the Islamic economists.

The course on contemporary Islamic economics is needed to keep the student fully posted on Islamic thinking in all the fields covered by the above mentioned subjects. There is a need for giving a synoptic view of whatever has been achieved till now by an Islamic critique of modern economics and by addressing Islamic principles to the actual problems of production, distribution, welfare, development, trade, international relations etc. A synoptic view will give a grasp of the unifying principles of the basic conceptual framework of Islamic economics.

Such a course at the graduate level pre-supposes that the student will be given a course on the source of Islamic guidance at the undergraduate level. This course should comprise the principles of Islamic jurisprudence and the techniques of understanding and interpreting the Quran and Sunnah. As far as possible the texts and precedents relating to economic affairs should form the raw material used.

In conclusion a summary of the graduate courses mentioned in this paper, is given below:

Core subjects (Compulsory)

1. Methodology of Economics
2. Microeconomic Analysis
3. Macroeconomic Analysis
4. Elementary Statistical Methods and Quantitative Techniques (Pre-requisite: An undergraduate course on Mathematics)
5. Public Economics
6. Economics of Growth and Developmental Planning
7. Comparative Economic Systems
8. International Economics
9. Structure of the (Home) Economy
10. Contemporary Islamic Economics (Pre-requisite: An undergraduate course on sources of Islamic Guidance)

These are all semester-length courses. Assuming two semesters each year, and four courses per semester, this leaves room for six optional courses. Alternatively, it may be advisable to extend the course on macro-economic analysis to two semesters as it includes money and monetary policy too. This will leave room for five courses. These may be selected from the following curriculum:

1. Quantitative Economics
2. Economic History

3. Area Studies
4. History of Economic Thought
5. Economics of Particular Industries
6. Agricultural Economics
7. Welfare Economics
8. Comparative Banking System
9. Financial Organizations
10. Labour Economics
11. Population Studies
12. Economics of Technology and Innovation

NOTES

1 Kahf, Monzer: *A Contribution to The Study of The economics of Islam.* University of *Utah*, SLC, U.S.A. 1973 110 p. (Mimeo)
and Siddiqi, M. N.: *Economic Enterprise in Islam*, Islamic Publications, Lahore. 1972 179 p.

Chapter Six
Beyond the Muslim Nation-States

Kalim Siddiqui

Kalim Siddiqui was a journalist in England and worked on *The Guardian* in Fleet Street. He holds the degrees of B.Sc.(Econ), M.Sc. and Ph.D. from the University of London. During 1972–74 he was Assistant Professor in the University of Southern California. He is now the Director-General of the Muslim Institute. Kalim Siddiqui's major works are *Conflict, Crisis and War in Pakistan* (London: Macmillan, New York: Praeger, 1972); *Towards a New Destiny* (London: The Open Press, 1974); and *The Functions of International Conflict* (Karachi: The Royal Book Company, 1975). He has also published numerous monographs.

In no field of human endeavour is the contemporary Muslim more confused than he is in the field of political science. This confusion is both at the intellectual level as well as at the level of the practitioners of the "art" or the "science" of politics — the politicians.

The confusion at the intellectual level is greatest among those least expected to be confused — the political scientists. The modern political scientist who is a Muslim is in great difficulty. He is a political scientist, with perhaps a doctorate in political science, a teaching post at a university, and even perhaps a few books to his credit. Yet the Muslim political scientist must ask himself a simple question: is he any different from non-Muslim political scientists who have identical degrees, university posts, and publications? The honest answer is "no" and that is also the correct answer. The catch lies in the phrase "Muslim political scientist." In point of fact "the Muslim" in the political scientist is independent of his academic discipline. There are, so to speak, two persons in one — a Muslim and a political scientist. The Muslim is the standard "believer" in Islam, but his political science is non-Muslim. The Muslim "faithful" and non-Muslim political scientist live in the single individual side by side and are the cause of much confusion. And when this schizophrenic "Muslim political scientist" sets out to pronounce on "the political theory of Islam" and "the Islamic State", the confusion is worse confounded.

The Roots of Political Science

It is not much more than fifty years ago that those paragons of wisdom, the professors of political science, were an unknown breed. The first of them was appointed in this century.[1] But when one asks what is the subject matter of politics, the immediate answer is the thoughts of Plato,[2] Aristotle, Augustine, Aquinas, Machiavelli, Dante, Hobbes, Locke, Rousseau, Bentham, Marx and John Stuart Mill. Then there are the descriptions of modern great States—the United States, Great Britain, France, Germany, the Soviet Union and a few others. Finally, there is the extensive contemporary literature on analytical conceptualizations.

If one looks at the above a little more closely, a number of questions arise. If, for instance, the "father of political science" (Plato) had written his *Republic* almost 2400 years ago, where has the child been in the meantime? The answer, partly, is that the child has been to Church for many hundreds of years and has then been put through the Reformation and the Renaissance. It then had to come through the faculties of law, history and philosophy before being recognized as a discipline on its own, though its twin sister, international relations, is still having some difficulty in being born.[3] This answer, however, is still not entirely satisfactory. The question still remains as to why did the Church and the faculties of law, history and philosophy hold the infant back for so long and then suddenly deliver it so quickly as a healthy child which has in fifty years (which is no time at all in the context of 2400 years) grown into a vigorous adult with a virile tendency to procreate? Why did this happen almost suddenly in the twentieth century? Why did it not happen in the eighteenth or nineteenth centuries or why could it not wait for the twenty-first century? Why, oh why, in the twentieth century? Why exactly at this time?

The answer(s) to this question holds the key to resolving a great number of mysteries.

One of the possible answers perhaps is that the political science we have was most *needed* at this time and hence its great success, great expansion, great recognition, great patronage and great following. Who needed it and why?

A. J. P. Taylor, the celebrated English historian, has recently written: "Europe took a long time to get going. Its lead [over non-European civilizations] began only in the sixteenth century [the Muslims lost Spain]; *its triumph came only in the twentieth.*"[4] Taylor was reviewing J. M. Roberts's book, *The Hutchinson History of the World.* Taylor becomes lyrical in Roberts's praise: "This is the unrivalled World History *for our day*.[5] It extends over all ages and all continents. It covers the experiences of ordinary men as well as chronicling the acts of men in power. It is unbelievably accurate in its facts and almost incontestable in its judgments." Taylor goes on to praise Roberts for holding "the balance fairly between the different civilizations" in his 1100 pages. Taylor then lets out the secret. ". . . he (Roberts) cannot resist devoting most attention to the European civilization he knows best *and to which he belongs.* Over half of his book deals with the recent centuries when Europe took the lead." Taylor does not regret this imbalance, but adds: "The reader will welcome this emphasis . . ." Why is Taylor so confident that the reader will welcome this emphasis on the European civilization? Obviously because Taylor knows that this is not history proper; this is the *Western view of history* and hence would be popular. This is indeed why he calls it *"history for our day"*.

Taylor thus tacitly admits that each civilization has to produce its own view of world history and other civilizations. No objectivity can therefore be attributed to his own judgments, let alone to those of his subject.

Let us stay with Taylor's view of history and accept, for the sake of argument, that Europe's lead began in the sixteenth century. He omits to tell us who was in the lead until the sixteenth century. We, the Muslims, happen to know, but the fact is carefully hidden from their readers by the historians of the West. Muslim civilization had remained dominant for over 1000 years and the Western civilization, as Taylor admits, did not in fact triumph until the twentieth century.

Once the triumph of the West had been finally accomplished and Muslims sacked from the stage of history, the West needed two types of intellectuals — the historians who would confine Islam and Muslims to a few paragraphs and footnotes, and political scientists who would justify and rationalize the dominance that had been achieved. In this enterprise to falsify history and produce a secular view of man and his political development, the newly triumphant civilization of the West proceeded to invest huge human and material resources. A third plank of the same strategy was the Orientalist tradition of scholarship, instituted largely to infiltrate the remaining body of Islam and to scuttle it from within. The Christian missionaries also joined in the same enterprise and received rich rewards.[6]

It is now possible to see clearly the roots of (Western) political science. These roots have not been allowed to spread of their own free will. They have not been allowed, for instance, to draw anything from Imam Al-Ghazali, Ibn Taymiyah or even Ibn Khaldun. Instead, the roots of modern political science have been carefully nursed to avoid contact with Muslims or Islam and to go directly to the ancient Greeks, the medieval Church, and back to feudal, and later national, Europe. Hence, right up to our own times, the political philosophers of the West are still arguing such issues as the nature of man and trying to explain political behaviour in terms of the Christian doctrine of "original sin". We have to bear in mind these contrived roots of political science. The Western political science, and Western history, philosophy and the arts, have all been contrived to serve the purposes of Western civilization.

One Essential Difference

The above analysis of the background to modern political science prompts a question: if each civilization needs its own political science, how did the Muslim civilization lasting over 1000 years manage without a political science of its own?

The answer is, in its own way, simple and yet complex. For Muslims generally, and for Muslim intellectuals and thinkers, political power and dominance was neither new and surprising nor did it need justification. For them, and for Muslim statesmen and administrators, political power was the very essence of Islam. They could not conceive of Islam or themselves outside the framework of a political system. To them political life was as natural as life itself; they took it for granted as they did sunshine, rain, air and the earth. They were like fish that did not have to stop to ask why water was necessary. Indeed, the *Sunnah* of the Prophet demanded the establishment of a political system without which Islam itself could not be understood or practised. There was no need to rationalize, theorize or explain.

So long as political power lasted and the political framework for the expansion and protection of the Muslim empire existed, Muslims worried little that the office of the *Khalifah* had become hereditary and, in essence, a monarchy. The ruler called himself *Khalifah* and, though he was no longer a selfless ruler, he was still recognized and obeyed as Amiru'l-Mu'minin.

This is in total contrast with the idea of the separation of Church and State in the history of Western political thought and development. Earliest Christians were organized, if they were organized at all, as a monastic order and not as political, military or civil units. Christians obeyed the Roman authority in virtually all matters. Ultimately, Church and State came into confrontation with boundary disputes between the profane and the sacred. Islam, on the other hand, began by defying the existing authority, by organizing civil and military and administrative systems, and, in the lifetime of the Prophet, defeating the opposition and establishing the unchallenged supremacy of the new way of Islam. As Iqbal puts it:

> "In Islam the spiritual and the temporal are not two distinct domains, and the nature of an act, however secular in its import, is determined by the attitude of mind with which the agent does it. It is the invisible mental background of the act which ultimately determines its character. An act is temporal or profane if it is done in a spirit of detachment from the infinite complexity of life behind it; it is spiritual if it is inspired by that complexity. In Islam it is the same reality which appears as Church looked at from one point of view, and State from another. It is not true to say that Church and State are two sides or facets of the same thing. Islam is a single unanalyzable reality which is one or the other as your point of view varies. The point is extremely far-reaching and a full elucidation of it will involve us in a highly philosophical discussion. Suffice it to say that this ancient mistake arose out of the bifurcation of the unity of man into two distinct and separate realities which somehow have a point of contact, but which are in essence opposed to each other. The truth, however, is that matter is spirit in space-time reference. The unity called man is body when you look at it as acting in regard to what we call the external world; it is mind or soul when you look at it as acting in regard to the ultimate aim and ideal of such acting. The essence of *Tawhid* as a working idea is equality, solidarity and freedom. The State, from the Islamic standpoint, is an endeavour to

transform these ideal principles into space-time forces, an aspiration to realize them in a definite human organization.'[7]

The above discussion and Iqbal's argument show clearly that the idea of State in Islam is fundamentally different from the idea of the modern nation-State. The two types of State are not the same thing. They have nothing in common. While Islam brings the State into existence as an instrument of Divine *purpose*, the nation-State comes into existence for precisely the opposite reason—to dismiss God and to replace Him with the "national-interest" as determined by human reason. Let us examine what one Western political scientist has to say. W. T. Jones agrees with Bodin that the concept of sovereignty was unknown to Greek or medieval thinkers, and goes on:

> "The reason is that certain conditions arose at the beginning of the early modern period which necessitated a new theoretical schema. The schema which was finally worked out is based on the notion of sovereignty, and, since the same conditions survive today, the notion of sovereignty is still of the first importance. These conditions are, of course, the emergence out of the feudal political system of the national territorial state. This kind of political organization had to come into being; or, rather, the modern world would not have developed as it has—into a lay, industrial and capitalistic culture—had it not been for the creation of the national territorial State, which is at once an instrument and an effect of this development ... They [Machiavelli, Bodin and Hobbes] see simply, but very clearly, that a supreme and absolutely powerful ruler—a sovereign—is an indispensable condition of the new order of things."[8, 9]

The Present Situation

We have now indicated some of the area covered by the conceptual jungle which is causing so much confusion in the political thinking of contemporary Muslims, including professional political scientists, politicians and also traditional *'ulama* attempting to analyse and prescribe for the political ills of the *Ummah*.

The confusion, incidentally, is at source—in the Western political science itself. There is, as yet, no agreement among the pundits of political science as to the meaning of "politics". Though the word "politics" originated with the Greek "polis", it has acquired many new shades of meaning.

Definitions of politics vary from "politics is a struggle for power" (Morgenthau), "the study of influence and the influential" (Lasswell), to "the authoritative allocation of values" (Easton).[10] Bernard Crick settles for the trite comment that "politics is politics".[11] Crick quotes Isaac D'Israeli as having said that politics is "the art of governing mankind by deceiving them".[12] The "dirty game" image of politics and politicians persists through the world, including the West. Some politicians, once they attained high office, have even attempted to put themselves "above politics". Others have tried to "save the country from politicians". Statesmen are often advised "not to play politics with the national interest". This is hardly surprising in a situation where the highest form of political organization, the nation-State, itself does not recognize any moral values, except those of the profane world. Such values are neither immutable nor universal. They are different from nation-State to nation-State, and in the same nation-State different values are often used in different situations depending on the "national interest" involved in each situation. Not surprisingly, therefore, international relations between nation-States represent a struggle for power, by means of power, for the sake of more power.[13]

Enough has been said to make the central point about the current situation in predominantly Muslim areas of the world today. It is simply that the Muslim World is now divided into nations and each nation has its own "nation-State". (Exceptions such as Palestine, Eritrea and Kashmir are under either foreign occupation or colonial rule). Muslim nation-States are essentially no different from all other nation-States. Few, if any, admit moral values, except as slogans. The fact, however, is that all nation-States are the product of Western civilization and its period of colonial dominance. Their purposes, structures and behaviour patterns are all alike whether the nation-State happens to be in Europe (its "mother" continent), or in Africa, Asia or America. The same analogy holds here as that we advanced in the case of the Muslim political scientist, where the individual was Muslim but his political science, "non-Muslim". In precisely the same way the countries are Muslim but their political structures — the nation-State — non-Muslim. Nationalism is the very antithesis of Islam. We must, therefore, face up to the situation and admit the reality that now stares us in the face: that no political manifestation of Islam exists today. Indeed, this phase of history when the Muslim populations are parcelled into nation-States is, strictly speaking, a continuation of the period of European colonialism. Instead of having direct rule from Europe, Muslim countries are now ruled by European institutions through local Muslim elites that share the European, secular and profane view of the world.[14] In a very real sense, then, we are all Europeans. And we are still under a form of Western colonialism. Indeed, the economics of Muslim "nation-States" are all capitalistic and integrated into the international capitalist economy. International trade, aid and finance and insurance corporations are the modern versions of the former East India Company.[15] It is in this context, the universalization of the nation-State and the capitalistic culture, that we should view A. J. P. Taylor's claim that the triumph of Western civilization came only in the twentieth century.[16] We are now "independent" prisoners of Western civilization.

Towards A New Political Science

Muslim political scientists must now talk as a group of prisoners. They must define the scale and model of the prison in which they live. They must map the prison in detail. The three dimensions of this prison are social, economic and political. These dimensions are linked by intellectual corridors of which the political scientists themselves are the leading exponents as well as its victims. To plan and ultimately execute an escape from this all-encompassing "open" prison, we may, for a while, have to behave like model prisoners and mix among our tormentors in a way that does not arouse their suspicion. To some extent it might even be possible to take the "guards" into our confidence. They might even co-operate with us so long as we do not become a threat to their positions and leadership roles in the short term.

We have got into this nightmarish situation through the cumulative effects of hundreds of years of neglect and the sins of commission and omission of our forefathers. There is, therefore, no responsibility on us to get out of this bog of history in one jump. The most we can do is build, or begin building, a solid platform from which a future generation can launch its escape. We have got into the present morass by default, but we can escape only by design. The first set of problems that confronts any team of design-engineers is conceptual. It is only after these have been cleared that model-builders can proceed with experimentation. Perhaps a period of experimentation and, hopefully, success will yield increased confidence and greater human and material resources for the final assault on the overriding forces of history. The first stage, however, is largely, if not exclusively, one of removing the conceptual obstacles and shaping a bagful of new conceptual tools. Whether the whole enterprise gets off the ground at all and proceeds to the subsequent stages depends largely on how well the initial tasks are accomplished.

It is clear that this first stage might be called one of "education". All educators must participate, whatever their field. Yet the political scientist has a central role to play. The Muslim historian does not need to write history as propaganda, nor does the Muslim political scientist need to theorize for the sake of appearing respectable and profound. In Islam there is a framework which bestows legitimacy on scholarship, especially teleological scholarship directed at goal attainment.

The goals the Muslim political scientist set himself have nothing to do with writing on the popular recent themes of "the political theory of Islam" and "the Islamic State". Such literature as was needed on these themes has already been written.[17] The goals that have to be set must be rooted in and derived from the present situation. As political scientists we know that the "nation-State" is alien to Muslim political culture and an importation from Europe. We also know that the present generation of Muslim nation-States have not solved and are unlikely to solve any of the problems that now confront the *Ummah*. We must prepare our students and future generations for the time when the nation-State will be no more. We must look beyond the nation-State and prepare the blueprint for a post-nation-State era. We must with our analysis prepare the intellectual climate which will look forward to the time when the nation-State will be no more. Some will go of their own volition; others might have to be brought down. The danger, however, is that the present institutions, bad though they are, might collapse before we and our peoples are ready with an alternative form of political organization to replace them. Politicians cannot be expected to plan to replace themselves and their systems; the Muslim political scientist has no choice.

The recent political "revivalist" movements among Muslims have failed to accomplish their desired goals. We need a number of dispassionate studies to try and discover why such movements as those of the *Ikhwan* and *Jama't-e-Islami* failed.

A number of possible reasons, or a combination of these, need to be examined. These must, of course, include their structures, leadership roles and "styles", and other "human" factors. But the greatest emphasis should be placed on the basic conceptual roots of these movements, their "reading" of the situation they tried to tackle and the policies they pursued. For instance, one would like to know whether the political party approach to change is acceptable. Was the Jama't justified, conceptually or on the basis of convenience and expediency, to jump to the conclusion that "elections" could produce the desired result? What is the place of expediency when the available options are in conflict with the values and traditions of Islam? Can an "Islamic movement" climb an unIslamic ladder and hope to arrive at Islam? What, if any "compromises" are possible or desirable in a "democratic" situation? Is the acceptance of the "democratic" process in a nationalist-capitalist framework justified?

Another range of questions that must be asked concern the social origins and relevance of recent Islamic movements. Were these movements too "middle class"? Did they represent, or appear to represent, established sectional interests in their society? Did they or did they not identify themselves with the poorest and the weakest in their society? Did the need for funds drive the Islamic movements into the arms or influence of those who wished to maintain the economic/social *status quo* under the banner of Islam? Did the Islamic movements appear to support the capitalists in the dialogue for change? What lessons are there for a future Islamic movement to learn from past, recent and current failures? Another set of questions concern the "level of competence" of the Islamic workers and the nature and extent of their commitment and life-style.[18]

Beyond the Muslim Nation-State

The Muslim teachers of political science face a most challenging task. Initially, they must reveal to their students and a wider public the true nature of the nation-State and all its structures and functions.

They have to develop a body of literature to prove that the nation-State cannot possibly solve any of the problems that now confront the *Ummah*. Contemporary history is full of data to drive this point home. The Muslim students of political science, indeed all students, must be made aware that they must look forward to and prepare for a time when the nation-State form of political organization will have disappeared.

But the passing of the nation-State, however desirable, must not be allowed to leave a vacuum or lead to disorder and anarchy, though some temporary imbalance during transition and fundamental change will be inevitable. The teacher of political science has to shape the mind of his Muslim students in a direction towards change. The political scientist, working among and with his students, has also to devise an overall strategy of change. He also has to produce operational models of change. The climate of opinion is to be so infused with the expectation of change that the coming of change will be welcomed and helped by Muslims everywhere.

Before that stage is reached, however, the Muslim political (economic and social) philosopher has to produce an image of the future which makes the present unbearable. A new set of socio-economic political systems of Islam have to be shaped. These models should be so rationally convincing and attractive that a whole new generation of Muslims would struggle to bring them into being.

This means that the Muslim political scientist must also develop in himself and his students the qualities of mind and spirit that would be required in the next phase of history. If the nation-State goes, with it must also go the present style, social origin and function of leadership. Perhaps we shall have to offer an entirely new Islamic concept of leadership. Maybe the word "leadership" would not apply to the active workers in an Islamic social order; perhaps all members of the new order would so participate normally and naturally in the promotion of the collective good that the "role" and "function" of leadership would become diffuse and institutionalized at all levels throughout the society.

It is, of course, impossible to foretell the future, but it would be negligent not to plan for it. It is also important to be conscious and realistic about the time-scale that would be involved. It would be useful, for instance, to divide the "future" into three parts. The short-term (next five years), the medium-term (next 20 years), and the long-term (20–30 years and beyond).

The processes of history are such that what is going to happen in the short-term is probably already beyond control or planning. The most that can be done over this distance is limited to marginal manoeuvres by top decision-makers. Few academics can possibly hope to influence events and their course over the short-term.

Over the medium-term the situation is not much better, though it improves as one gets into the latter part of the period. Though "events" might still exercise a strong momentum of their own, it is possible to influence our "response" to them. For instance, it may be possible to make our social, economic and political systems aware of a wider range of alternatives in determining their behaviour pattern. This would greatly improve the efficiency of these systems and the output per unit of resources might also improve dramatically. But in the 20-30 year range it would be possible to project a period of intense activity for the attainment of major intermediate goals; these goals would be the type which might be called "pre-requisites" for the ultimate triumph of the "Islamic Movement" over all other forces, internal and external. Just what form this triumph would or should take is itself a challenge to social scientists of the present time.

What is not in doubt and incontrovertible is that only a major act of will and long-term planning can deliver the *Ummah* into the next phase of history that lies beyond the Muslim nation-State.

NOTES

1. W. J. M. Mackenzie, *Politics and Social Science*, London: Pelican, 1967, p. 57.
2. Plato is given the title of "father of political science" and his book, *Republic*, is the first book of politics.
3. J. W. Burton, *International Relations; A General Theory*, Cambridge: The University Press, 1965, Chapters 1 & 2.
4. The *Observer*, London, 19 Dec. 1976
5. My emphases throughout.
6. Christians now admit that for more than a century "many missionaries served the interests of the colonial powers." See the "Statement of the Conference on Christian Mission and Islamic Da'wah", Chambesy, June 1976, published in *Impact*, 6:21, 12-25 November 1976.
7. Muhammad Iqbal, *The Reconstruction of Religious Thought in Islam*, Lahore: Ashraf, 1971, p. 154.
8. W. T. Jones (ed.), *Masters of Political Thought*, vol. II, London: George G. Harrap, 1963, p. 19.
9. For a discussion of the origins of the nation-State system in Europe, its

universalization through colonialism, and its impact on non-European areas of the world, see Kalim Siddiqui, *The Functions of International Conflict*, Karachi: Royal Book Company, 1975, especially Introduction and Chapter II, "Political Legitimacy in the Third World". See, also, R. Emerson, *From Empire to Nation*, Boston: Beacon Press, 1962.

10. Ibid., and also see Kalim Siddiqui, "Is Politics Relevant" in *Impact*, London 3:5, 27 July–9 August 1973.

11. Bernard Crick, *In Defence of Politics*, London: Pelican, 1964, p. 16.

12. Ibid.

13. The leading exponent of this view of international relations is Hans J. Morgenthau. See his *Politics Among Nations*, New York, Knopf, 1948 (Fifth edition 1973).

14. It was this view that led me to propose a model of the "Islamic Movement" in which Muslim States are merely sub-systems, see Kalim Siddiqui, *The Islamic Movement: A Systems Approach*, Slough: The Muslim Institute, 1976. This paper is also an example of how an analytical tool of modern political science might be used by a Muslim political scientist.

15. It is this thesis that has been treated at length in a study of Pakistan. See, Kalim Siddiqui, *Conflict, Crisis and War in Pakistan*, London: Macmillan, and New York: Praeger, 1972.

16. Loc. cit.

17. I am not convinced of the efficacy of this literature. I find it apologetic in tone. Authors who have written in this framework include such eminent figures as Maulana A. A. Maudoodi. It appears to me that these authors have tried to mould Islam into the framework of Western political science, they ask who is the sovereign and come up with the answer "Allah". They then spend pages on the concept of sovereignty in Islam, though the fact perhaps is that the concept as such never occurred to Muslims or Muslim rulers. But I am not qualified enough to challenge these authors and their literature. I am therefore confining my unease to this small footnote.

18. This point is discussed in my paper *The Islamic Movement: A Systems Approach*, op. cit.

Chapter Seven
The Ummah and its Civilizational Crisis

Abdul Hamid Abu Suleiman

Abdul Hamid Abu Suleiman. Early education Mecca; B. Com. (Al-Azhar), 1959, M. Com. (Al-Azhar) 1962; Doctorate from the University of Pennsylvania, Philadelphia, USA, 1973; Assistant Professor, Faculty of Commerce, University of Riyadh; Formerly Secretary-General World Assembly of Muslim Youth, Riyadh, Saudi Arabia.
Translated from the Arabic by Isma'il R. Al-Faruqi.

I. *In Search of the Proper Avenue*

Despite all their problems, Muslims still claim to belong to Islam. However, neither their claim nor Islam itself may be taken seriously as long as Muslims continue to suffer from numerous shortcomings of which Islam is the clear contradiction.

Muslims number at least one fifth of the population of the world. For centuries, Islam dominated in religion and culture. It has been the object of countless attacks by its enemies; and Muslims have been subject to all kinds of aggression because of their adherence to it. And yet, Muslims cling tenaciously to their religion. This fact notwithstanding, many Muslims have turned to many recourses to exchange the ideology, which Islam gave them, for a new one. They untiringly imitated the non-Muslims, whether from the East or the West. They adopted their laws, customs and methods piecemeal or wholesale, and for long periods of time. In some cases, as in Egypt, imitation of the West continued for nearly two centuries. All of this was undertaken with the purpose of changing their situation. But no graft, no transplant and no imitation of East or West ever succeeded. Furthermore, the proposals for replacement of Islamic ideology were not viable; indeed were proven unfit. Humanity everywhere is not only dissatisfied with Western ideology despite its great scientific and technological advances, but is definitely deteriorating and retreating from psychological, emotional and spiritual stability and contentment. This so-called "modern civilization" has so far failed to provide man with the basic metaphysical premises capable of bringing the awesome powers it has unleashed to work for man's perpetual bliss rather than destruction.

All these facts compel a Muslim search to go beyond the West, to outgrow the stage of imitation. How else is he to prescribe a cure when those who are at the apex of modernism are falling to the ground in instability and malcontentment? However, the Muslim World must negotiate with Western ideology and its adherents. Isolationism is no longer possible or beneficial. A significant part of humanity, the Muslims — and hence, their problem — are inseparable from the larger problem of mankind, which finds itself in a world divided against itself, maintained by a balance of terror, and pressing headlong toward disaster.

More specifically, the problem of the Muslim peoples at the end of the fourteenth century after the Hijrah shows the following grave aspects:

1. Muslim deterioration is not a modern, but rather an old phenomenon whose roots reach beyond the fall of Baghdad, Granada or Samarquand.

2. All the catastrophes and disasters which afflicted the Muslims in modern times have not succeeded in destroying or changing them, or even causing them to alter their course.

3. All the past and continuing efforts inspired by non-Islamic sources which have been brought forth to solve the problem of Muslims have failed to move or mobilize the *ummah* to support them.

4. The main principles of modern Western civilization have in fact led humankind to tragedy despite all its positive achievements. A future entrusted to them will inevitably be a dark one.

Such grave aspects compel the most serious examination. They demand a probing in depth of the very existence of the Muslims as subjects of history. We must seek to uncover both the Muslims' potential and efficacy on both material and spiritual levels if their crisis is to be correctly diagnosed and successfully solved. Our first question is whether the crisis is one of the Muslims' material potential and the answer is certainly "no". Muslims inhabit and dominate a wide portion of the globe, an area capable of providing them with all their material needs. They belong to all the ethnic and cultural groups of the world and in such large numbers that their cultural poise and self-confidence are well established. Their past civilization and achievements give them further grounds for reassurance. And yet, their material poverty, inefficiency and impotence to provide their own most basic needs make them the beggars of the world, the sick men of the world, and made of their lands the trouble-spots of the world. Their shortcomings on the material level are not those "given" of nature but in consequence of their moral and spiritual malaise. Hence, their problem ought to be properly diagnosed on this level.

Students of Islam, observers of Muslims and monitors of their history, have offered views of Muslims' spiritual malaise. Most of them have pointed their finger at Islam — its religion, law and ethic, the culture of the *ummah* — and charge it with responsibility for Muslim decay and retrogression. Some Muslims have listened to this advice and acted upon it. The last two centuries saw their efforts to apply this radical solution of supplanting and exchanging Islamic institutions and laws for those of the West. In some countries, these "reforms" were carried out slowly and piecemeal; in others, such as Turkey, wholesale and with surgical speed. And yet, none succeeded in altering the *ummah*'s downward trend. Everywhere Muslim problems seemed to increase, until the vision of its leaders everywhere is currently dimmer and more out of touch with reality than it has ever been.

Evidently, Muslim leaders have missed their objective. They have erroneously sought it in the fundamental doctrines of the *ummah*, in the faith of Islam, the values and ultimate goals embodied in its laws and ethic. It is common knowledge that the *ummah* fell short of these values and goals; but it is not at all evident that the cause of their shortcomings is in the values and goals themselves; the values of Islam are beyond reproach. Only cynicism and malevolence may move one to attack them, a motivation not beyond orientalists, missionaries and other enemies of Islam. Instead we should search for the cause of the shortcoming in Muslims themselves even though Muslims are not ready to repudiate the values of Islam.

Before continuing the search this question regarding the values of Islam must be raised and answered once and for all e.g. Do the doctrines, first principles, values and social goals of Islam as such present any weakness or discrepancy? Who would not like to see them realized and fulfilled? Has any Muslim really lost his faith in them? Or is it once more purely the Muslims' actualization of them that is faulty? Are honesty, trustworthiness, mercy, self-exertion, mutual solidarity, justice and perfectionism questionable values? Does not the malaise lie in Muslims' failure to abide by these virtues? Are they not desirable — even necessary — for facing the crisis? Would not the crisis be more critical had the Muslims completely forsaken and abnegated these values?

What then is the exact nature of the problem? If it is not pertinent to the values in themselves, nor to the material circumstances in which Muslims live, must it not rest in their implementation? Implementation has two components, vision (or thought and planning) and action. Again, the possibilities of action on the part of a billion souls living in the temperate belt of the world are infinite. The malaise, therefore, must be in the Muslims' vision. It must affect their perception, understanding, appreciation and apprehension of the values of Islam. It must lie in the Muslim's capacity to be moved by the moral imperative, his sensitivity to the moving appeal of Islam.

II. *Analysing the Problem*

Beyond doubt the earliest period of Islamic history, the period of the Prophet (SAAS) and his Rashidun successors in Madinah, 1–40 A.H./622–659 A.C., constitutes the base of the Islamic edifice in history. The achievements of that period were extraordinary on all fronts, in all fields. Muslims in every age looked to them for inspiration and norms. No student of history fails to observe that a shift did take place at the end of that period, and one which set it apart from all that followed. And yet, hardly anybody has stopped to ask, Why? What is the nature of the shift that took place? What were its causes? Everyone has noticed the consequences of the shift; few have sought to penetrate and explain it, and none has seen in it the cause of the Muslim retrogression that followed. The reason is that the student of that period of Islamic history is dazzled by its *éclat* and splendour, victories and exploits, material wealth and moral abundance. It is extremely difficult to relate any retrogression or decay to these outstanding successes which continued well into the Umawi and first Abbasi periods. Even when incidents of political coercion or terror are pointed out, the observer is likely to dismiss them when set against the overwhelming evidence of cultural and civilizational excellence; or to bury their memory in compassion for the contemporary tragedies of the *ummah*.

The early period requires more analysis than our historians have so far given it. For it is in these very events in the early history of the *ummah* that the cause of later decay are to be found. The malaise which has persisted so long and resisted every remedy cannot be properly diagnosed and treated unless its origins are traced deep in the *ummah*. There, right at the fountainhead of that tremendous expansion of the *ummah*, must also lie the germ of decay which grew and followed its own course despite the expansion and flowering of civilization around it.

As the wave of Islam began to spread across the Near East, the inhabitants of East, West, North Central and South Arabia rallied in great numbers to the call of Islam. The *Muhajirun* of Makkah (the Quraysh) and the *Ansar* of Madinah (al Aws and Khazraj tribes) were in the minority and the other tribes preponderant. Certainly the two groups were not of the same Islamic species which the Prophet (SAAS) had moulded and disciplined, educated and prepared, to whom he had taught the Islamic vision and to whom he had imparted the fire. The majority of tribesmen were not privileged to be among the Prophet's disciples, nor did they participate in the building of the first Islamic polity under the Prophet's leadership (SAAS). As the Quran had spoken of them harshly but truly, "the tribesmen are stronger in ungodliness and hypocrisy. It can only be expected of them to ignore the laws Allah has revealed to His Prophet . . . The tribesmen claimed that they believed. Say to them (O Muhammad): 'Do not make such a claim; and say only that you have acquiesced. For *iman* (conviction) has not yet penetrated your hearts and minds' " (9:97; 49:13). The harsh environment in which they lived permitted little beside survival, and Islam was still too new to have become second nature to them.

Having formed the armies which conquered the two empires of Persia and Byzantium, it was also necessary that the tribesmen should make their strength felt in the decision-making of the *ummah*. The *Muhajirun* and *Ansar* were declining, older, and perhaps a little weaker on that account. Their influence could not be exercised without confrontation with the tribesmsn, which nobody would have wished or permitted. This influence was mainly responsible for the transformation of the Prophet's policy of Madinah into the Umawi imperium. A few daring Muslim historians have in fact spoken of this political deterioration, but they remained silent on its cause, assuming that the early fathers could have committed no mistake. Indeed, those historians who did notice the political transformation, spoke of it as if it was a phenomenon *sui generis*. It had not occurred to them that political transformation was the index of a wider, and more significant change.

Accompanying the political events which led to the rise of the Umawi regime, was the separation of the state leadership from the thought and vision of Madinah represented by the remaining companions of the Prophet. Abu Dharr al Ghifari (RAA) embodied that vision of Madinah when he stood up to answer Muawiyah, the state leader, as the latter sermonized from the Prophet's pulpit. Commenting on the conquests, Abu Dharr said: "All wealth, especially and including the new-conquered wealth, belongs to Allah." The tribal view was embodied in Muawiyah who utilized the new wealth to build up a party, to appease enemies and win friends—all to the end of consolidating his grip on political power. In this spirit, he declared, in contradiction to Abu Dharr: "Rather the wealth belongs to the Muslims!" The Madinese stood fast behind their vision and opposed the new Muslim "establishment" in Damascus even though Muawiyah belonged to the Quraysh tribe. The confrontation led to the revolts of al Husayn, Ibn al Zubayr, Muhammad al Nafs al Zakiyyah and Zayd ibn Ali, all of which were brutally crushed. Practically all great thinkers of Islam in the first century and a half following the Rashidun caliphate were alienated from the state leadership and sympathetic to the opposition. Imam Abu Hanifah died in jail; Imam Malik was subjected to public beatings; Imam Shafti'i was compelled to flee, and Imam Ahmad was tortured in prison.

Of all the features of history of that early period, the political division and strife between the parties was the least significant; the separation of the political leadership from the spiritual being more important and basic. Its implication is that thought had alienated itself from action, or *vice versa*. Without the Islamic vision, action by Muslims loses the Islamicity which association with thought confers upon it. Thought, for its part, dispossessed of its natural sources, weakened and began to dry up as its waters receded. Its isolation deprived it of challenge and the chance to renew itself. Politics, on the other hand, deprived of thought to Islamize, justify and (where necessary) redress it, deviated from the straight path. The removal of moral and spiritual values deprived political action of guidance, and inclined it toward coercion and reliance upon ignorance or indifference to the masses. Action thus brutalized itself through separation from thought. Certainly Muslims continued to act strongly and to plan; but their actions were expedient, their planning vain. In this separation lay the greatest danger to the future of Islam on earth, the very source of its tragedy and downturn of its *ummah*.

All later Islamic history was affected by this split between thought, or Islamic vision, and action. The rise and fall of Muslim states everywhere — Damascus, Baghdad, Cairo, Cordoba, Aleppo, Delhi and Istanbul — were ultimately related to it. The poor vision, the tribalizing of pre-Islam, the Persianisms, Hinduisms, Byzantinisms, Israelitisms and Christianisms which penetrated almost all fields of Islamic thought and practice — followed in consequence of that original separation of the two leaderships. Whereas the rulers ruled by force, the thinkers sought to establish and confirm the Islamic vision in consciousness. To this purpose they first developed the sciences of *hadith* and law, with their attendant descriptive disciplines. The thinkers began to elaborate superstructures of speculative doctrine removed from the social system and its practices. Their greatest energies were wasted on problems of transcendent nature such as the divine attributes, free will and determinism, and creativity of the Quran. The masses, for their part, devoid of leaders to protect them from the deviant state authority, surrendered meekly to their rulers and to myth and superstition.

Furthermore the self-isolation of Islamic thinkers made them less capable of understanding events around them. Distance from events prevented the thinkers from containing them, or in any way influencing their course. Fearful of the arbitrary rulers who might undermine the genuine Islamic heritage, the thinkers advocated conservatism and literalism. They contented themselves with their personal faith and conviction, but became more detached from the affairs of state and society. Finally they declared any creative interpretation of the law (*ijtihad*) as forbidden once and for all. The limited creativity which subsequently appeared in Muslim history came from those exceptional leaders like Imam Ibn Taymiyyah who dared enter the political and social arena. Creative reformative thought must spring from the hotbed of practice if it is to address itself successfully to that which needs to be reformed. Consider in this case the creative interpretations of Umar ibn al Khattab (RAA), Amir al Muminin, which he injected into the understanding of those Muslims who stood in such close proximity to the Prophet (SAAS). He ruled that the conquered lands of Iraq belonged to their non-Muslim owners, who must henceforth pay a *"kharaj"* tax to the public treasury for dispensing, to maintain the general welfare. This ruling, which ran counter to the practice of the Prophet (SAAS), was born in the very act of conquest and possession of these lands by Muslims. Another ruling of his — that the three declarations of divorce pronounced at once counted as three separate ones and made a divorce final — ran counter to the Quranic pronouncement itself (2:29). Umar, however, was concerned with the unfair treatment in the new provinces which women were receiving at the hands of their husbands, and sought to make divorce more important, more serious and consequential than Muslims were inclined to regard it.

From all this it can be seen that the methods of Islamic thought to which Muslims have been accustomed are in need of radical reform. Starting from the overall vision i.e. the general objectives and values of Islam we need, equally, action to revise, reform, and change its practice. Only then will Islamic thought regain its realism and vitality and be again united to life and history as it was in the early period. Such radical reform presents Muslims with a great responsibility. They must learn to look at the events of the past with the eye of the keen historian who seeks to uncover causes and contexts of phenomena, to relate these to the objectives of their agents, and to establish the flow of history as a strong chain of linkages of ideas and actions. We must learn again to understand our past as if we were participating in it at that time. Only then will the opaqueness and arbitrariness of past events give way to clear vision and understanding. Only then will Muslim thought free itself from its shackles and assume its rightful place as guide to Islamic life and action.

Without this radical reorientation, Muslims will continue to move in vicious circles, and their tragedies will continue as victims to powerful nations.

III. *Toward a Solution*

Such reorientation as we have so far advocated is of course a political decision of the gravest kind. In the main, it consists of recasting "education" and "public information" anew, of relating them to an integrated culture and style of life. Islamic education and information ought to become the basis of a new Islamic leadership; ever-conscious of the Islamic vision; committed to its realization in history, and *engagé* in the *ummah* as its fundamental source. Leaders and *ummah* must be bound together in an inseverable bond of consciousness of destiny and activist pursuit of its content. Education must present its programme with the objective of preparing men with the competence to interpret ideas, facts and events, in the light of the ultimate values of Islam, and to do so on the highest level of exacting scholarship. Public information, on the other hand, must feed the masses with precepts of the values of Islam in all fields of human endeavour, and do so creatively in a manner designed to inspire and move its audience. It is the duty of the Islamic thinker, as well as of the leaders of the contemporary Islamic movement, to specify in the clearest terms the nature of the decision required of the *ummah*. The *ummah* remains responsible for the fulfilment of the decision; for the realization of its content by all means, and in all departments of life and activity. Both thinker-leader and *ummah* should form an inseparable unity which will reflect the interdependence of the organs of a living body. Public and private education, public instruction and parental instruction in the home, employment policies and policies of foreign relations, measures of social welfare and rules of judicial procedure — all should reflect the one "Straight Path" of Islam. Then and only then, Muslims could say that the *ummah* is on the march again, on that path which the Prophet (SAAS) had blazed for it.

Any reform should endeavour to correct the *ummah*'s concepts and categories of judgement, above all past prejudices and traditional misinterpretations. These belong to a system we are seeking to outgrow. To this end, our outstanding scholars ought to mobilize study, and prepare to review the *ummah*'s concepts of itself, its religion, history and mission. This critical examination should be carried out in all fields, so that the relevance of Islam in each may be clearly established and understood by specialists and workers alike. Once the relation of any given event to the Islamic value of which it is the spatio-temporal carrier is appropriated by the mind, one could relive that event in the imagination and re-enact it as if one were its actual subject in history. It is through such comprehension that we can train people to be — like the Prophet's Companions (RAA) — men and women in constant motion, working and inspiring others to work for the good of the Islamic cause. For the Prophet's Companions (RAA), Islam was not a moment of *dhikr*, nor one of charity, but a persistent advance in the cause of Allah. They did not feel the poorer for meeting their fellows with the smile of welcome, nor for stooping to remove the smallest obstruction from the public highway, as the Prophet (SAAS) had directed. After all, the merciful have made themselves worthy of Allah's mercy.

Such people genuinely believe in the following precepts of the Prophet (SAAS), and consistently think and judge, as well as decide and act by them. They are the universal laws taught us by the Hadith:

> "The Muslim is he from whose hand and tongue all humans are safe ... whose evil influences never reach the neighbour ... The Muslim is the brother of every Muslim; never does he betray or harm ... Allah will help man as long as man helps his brother ... The Muslim is to the Muslim like one part of a solid wall to another, each supporting and consolidating the other parts ... In their mutual affection and compassion, the Muslims are like the body: When one organ falls ill, the whole body rises to its assistance ... No superiority for any Arab over a non-Arab unless it be by virtue or piety and righteousness ... That others make themselves guilty of injustice and aggression is no excuse for you to do the same. Therefore, be always just; that is closer to piety — Quran 60:8. Righteousness is nothing but in the deed ... He is no believer who retires in any night knowing that his neighbour is hungry ... Cleanliness is a very constituent of faith ... Whatever it is that you may do, Allah wants you to do it well."

A few thousand people who truly believed in these values, who were inspired by them and spent their lives in applying them, conquered the "civilized" world of their day. They won the greater part of humankind to their faith. Everywhere people modelled their lives, their character, even their language on them. History has never known conquerors so successful in such a short time. Their very conception of religious duty was never restricted to the personal dimension, even when their duty was related purely to Allah i.e., in adoration. Their very worship was inseparable from their altruistic deeds which always had a positive element. Truly, they saw themselves as servants of Allah, on a par with all His human creatures; and in their service to Him found dignity and meaning in their lives.

The real nature of the Prophet's Companions (RAA) is evident in another chapter of the history of that period, that of the Riddah War under Caliph Abu Bakr (RAA). Students of history and political theory have never succeeded in correctly understanding the cause and meaning of Abu Bakr's decision. Noting the incomplete political structure of Muslim society and the preoccupation of most of its leaders with foreign conquests, they conclude that the Companion's conduct does not provide a model worthy of emulation. Their failure to grasp the deeper implications, that moved Abu Bakr and his colleagues, causes them to see the Riddah events as merely political.

Abu Bakr (RAA) was the best of the Prophet's Companions. He was chosen unanimously to lead the *ummah* at the death of the Prophet (SAAS). He was known for his tenderness, mercy and deep compassion. He often wept out of sympathy; but his vision was crystal clear and his will unshakable. When tribesmen rebelled against his authority and stopped the payment of *zakat* to the central authority of the Islamic state, the question for him was not one of faith. The Quran had previously described them as having acquiesced to, but not as convinced of, the Islamic truth. It is up to Allah to settle such matters on the Day of Judgment. No more need be added. For Abu Bakr, the question concerned the social system; of man's relation to man as the essence of righteousness, and hence, of Islam. To deny the social system as those tribesmen had done was to deny Islam and so warrant the Islamic State's punishment.

The Prophet's Companions raised their voices in indignation and opposition. The Prophet was hardly buried, when the Muslim army was committed to fight the Byzantines, who were amassing their forces in the North, and Abu Bakr wanted to fight his fellow-Muslims in Arabia! Abu Bakr stood firm and defended his decision. He must either convince the Companions of its propriety or they convince him of the opposite. Dialogue began, for Abu Bakr would not continue in office without both the support of the Companions and the knowledge of a righteous decision in a matter such as this. Slowly but surely, the Companions were won over to Abu Bakr's conviction that Islam is not merely a question of personal faith and conduct. It is a social order of which the *zakat* and the central governmental authority are components. Whatever the quality of a Muslim's personal faith, he may not betray the *ummah*, the ideological society set in motion by the Prophet. Umar ibn al Khattab (RAA) who held at first the opposite view, announced: "And when I beheld the gravity of Abu Bakr's commitment to the matter and understood its logic, Allah enlightened my heart, and my mind inclined to agree with him". Thus Abu Bakr fulfilled his duty as leader of the *ummah*, and brought the Riddah War to a victorious conclusion, not in opposition to the Companions, through coercion and terror, but with their wholehearted support.

The Riddah War is not a case in which the *shura* principle has been violated. To believe to the contrary would be evidence of ignorance of Islam and its decision-making process, for *shura* and the democratic principle are not one and the same. They stem from different social philosophies, and ultimately form a different representation, or understanding of reality. Democracy bases itself upon a materialist, individualist theory under which the individual enjoys an absolute right to express himself as an individual. His political inclination has an absolute value, unassailable because it is his, regardless of its content, or of its consequences upon the destiny of society. Indeed, according to his view, society itself exists to enable the individual to express himself. The individual's exercise of his right is the supreme norm, as the individual's personal life is the ultimate base. Hence, when individual persons form a majority group, their decision, as right of the majority, is sacrosanct. Even the constitution is only an effect of such a majority, alterable as its will. That is why in most democracies, any arguments questioning the truthfulness of the constitution, or criticizing it in terms of Allah, or of any other authority above the majority's decision, are *ipso facto* illegitimate, since there is no law above what the majority had decided to be law. In this system, the minority ultimately has no rights whatever; and the constitutions of modern states and bills of human rights are only patchwork systems at the service of the majority.

This is all a far cry from the *shura* principle which is an expression of human brotherhood. *Shura* is the principle of individual representation designed to fulfil the social order of human brotherhood in accordance with Islamic law. Its *raison d'être* and spirit, its end and meaning, are never lost sight of — let alone violated — in the implementation. The individual's right, as well as that of the majority, are indeed respected, but only within the bounds of the Sharia. They are not absolute, not even when their agreement makes up the constitution of the state. Individuals are required to exercise their rights in the spirit of brotherhood, and for the benefit of society as a whole. The body politic is not to be divided into minority and majority, and its decisions must serve the totality and do so in the style of human intercourse peculiar to Islam. The decisions must be the fulfilment of truth and values, not an arbitrary self-expression, or an exercise of the absolute right of the individual as such. They must bear out the Islamic values of mutual consultation and advice. Allah — *Subhanahu wa ta'ālā* — has commanded: "Uphold justice, and witness unto Allah alone, even if this goes against your own persons, or your parents and relatives. Do not follow your personal passion, lest you violate justice" (Quran 4:135).

It is this kind of intensive effort, capable of correcting the categories and concepts of education, of social conduct and group activity, that is now needed to reform education and information among us. It is the solution to the "crisis of thought", the alienation of the intellectual and actional leadership which afflicts the *ummah*. It will restore to us our original unity of culture and enable us to outgrow our centuries-long retrogression, weakness and decay.

IV. *Relevance to the World*

Finally, it may be asked, what is the relevance to the world of the present crisis of the *ummah* and of its possible outgrowth?

A. The first principle pertinent to this matter is the design Allah has drawn for human relations. His will, crystallized in the Islamic revelation, is for a society built on a foundation of unity and brotherhood, mobilized to the goal of satisfying the basic needs of the individual as well as of society. Islam has indeed provided means of regulating the social order at each level of society.

It is not viable for nations today to live in constant confrontation with one another, under a system of "balance of power" or "balance of terror". Nor is it possible for any to live peacefully if its population suffers from class conflict and internal strife. The ready availability of modern weapons, the atmosphere of competition and conflict, and the absence of over-riding factors in terms of which a conflict may be solved, might well lead to world destruction. The system of mutual solidarity and interdependent links which Islam proposes as a design for human society are inseparable. Islamic philosophy guarantees growth of inner spiritual and emotional contentment for all humans. The principle of solidarity and interdependence is the political and social figurization which the following divine verses imply:

Allah — *Subhānahu wa ta 'ālā* — said in His Holy Book:
"O People! Fear your Lord! Revere Him Who created you of a single person; Who created from that person its spouse; and from both, so many men and women . . ." (4:1).

"O People! We have created you all of one male and female; and We have constituted you into tribes and nations to the purpose of knowing and cooperating with one another. Nobler among you is the more righteous. God knows all things" (49:13).

"It is of the signs of Allah that He created the heavens and the earth, and made your tongues and skins diverse" (30:22).

"Humans were not but a single *ummah*. Only later did they differ and divide . . ." (30:22).

"Allah commands kindness and good deeds to parents, to the relative, the orphan, the deprived, the neighbour whether relative or stranger, the companion by your side, the wayfarer . . ." (4:36).

"For this reason, We imposed upon Banū Isrā'īl the law that whoever unjustly kills a person or spreads corruption and mischief on earth, must be treated as if he had killed humankind; that whoever saves the life of one person has saved the whole of humankind" (5:33).

"Never omit to do good to one another" (3:237).

". . . And always have kind words to say to people" (3:83).

Allah does not prohibit you from doing good and acting justly and fairly to people who have not fought against your religion, nor expelled you from your homes. Allah loves all who do justice" (60:9).

"And if you have to punish, then impose exactly the same punishment as you had suffered. But if you suffer with patience, it is certainly better" (16:126).

"Fight in the cause of God those who fight you. But do not transgress" (2:190).

"If the enemy stops fighting, there shall be no aggression except against the unjust" (2:193).

"The unrighteous deeds of a people should in no way be an excuse for you to commit injustice. Pursue justice always! That is nearer to Allah" (60:8).

"And if you have to speak up, do so justly, even if it were against a person of relation to you" (6:152).

'Whenever you adjudicate between people, do so with justice and equity" (4:58).

"Cooperate with one another for goodness and righteousness, not for crime or aggression" (5:2).

"If two groups of believers fight each other, reconcile them. In case one group violates the accord, then fight that group until it returns to Allah's commandment. When it does, reconcile them again with justice and equity. Always do justice. Allah loves the just. The believers constitute but one brotherhood. Reconcile, therefore, your brothers to one another. Fear God, that He may show you mercy.

O Believers, do not permit a people to ridicule or despise another people, or women to hold other women in contempt. The despised may be better than the despiser. Do not defame one another, nor call one another by offensive nicknames . . ." 49:8–10).

B. The second principle concerns knowledge and the methodology of researching it. Allah has revealed it for the purpose of helping man to build the ideal social order. In the main the civilizations of the world and the tradition of materialist thought have developed on a foundation of rationalism, empiricism, and induction from fact and experience. From the world of the senses, of observation and measurement and from the cumulative experience of mankind, man drew out his knowledge of the laws and patterns which govern life and the cosmos. This knowledge has stood apart from that which revelation has brought to all the world religions except Islam. The best known case of such disparateness between revelation and the findings of reason is that of Christianity, where rational and scientific knowledge ran counter to Biblical scripture. The only case where such disparatness is utterly absent is Islam, where reason and revelation are equivalent and united in the Islamic doctrine of the unity of truth as a facet of the unity of Allah. The Islamic faith recognizes no possibility of a contradiction between revelation and reason. Reason affects the observation, classification, measurement and reporting of data; revelation the understanding of the texts of revelation, or the reasoning derived from such understanding. Revelation is thus regarded in Islam as immune against any attack. Islam receives conflicting claims for further study and analysis, certain that the said claims can be reconciled under a discoverable principle over-riding both. For this reason, Islamic history knows of no Brunos or Galileos; and Islam has no "church" endowed with magisterium, no person or institution capable of making *ex cathedra* pronouncements.

On the other hand, humanity needs revelation to provide it with immutable, eternal laws for guidance. The nature of human society, or social relations, is so complex, and the factors influencing human conduct are so diverse and intractable, that any restriction of social reality to a so-called "scientific" theory is almost certain to be false and result in misinterpretation of other facts. Nor is it possible to put society in a laboratory for a completely controlled experiment testing the veracity of hypotheses. That is why the present state of knowledge of social reality among the so-called "social scientists" is chaotic. Theory after theory is raised and defended with fanatical persistence by its proponents one day, only to be contradicted by a newer theory and regarded with contempt on the next. All Western disciplines of social behaviour and education reflect the same degree of confusion and untrustworthiness. Already, the mistakes they have committed have left their adverse – oft sinister – effect upon Western societies. It will take incalculable effort to cleanse society of their evils.

It is otherwise with the tradition of Islamic knowledge and learning. Revelation includes all knowledge of sciences. In so far as that knowledge is true and is a reflection of reality, it is accepted and blessed. No suspicion may hamper such acceptance, if the knowledge is true. On the contrary, Islam deems the pursuit of such knowledge desirable, indeed obligatory upon all men and women. Allah has made the whole of creation subservient to man, and commanded man to use it, to make its fruits available to all. Obviously this necessitates mastery and manipulation of the elements and forces of nature and, in turn, a real and complete knowledge of them. Such is knowledge of the immutable patterns of Allah which are none other than the real laws of nature.

The strength of Islamic knowledge is that, having observed the conclusions of science, it presses far beyond them to *a priori* principles revealed by Allah for humanity's benefit. These concern the basic social conduct of mankind. The Muslim social scientist may carry human inquiry as far as he wishes. If he there discovers the divine patterns and establishes them critically for the ready use of understanding, he has earned merit. But if he is uncertain about his conclusions, he must have recourse to the treasury of the truth of Islamic revelation, none of which runs contrary to reason. This recourse constitutes for him an added reasonable guarantee, of refuge, not contrary to common sense.

The Muslim social scientist's infallible guidance is the Islamic principle that nothing which opposes a clear textually established permission or prohibition is true; and any that does not so oppose is innocent, permissible and probably true. The Muslim may enter any activity — and the Muslim social scientist must bless such entry — as long as no harm or injustice befalls him or others in the process. The Muslim may indeed turn his personal and family life toward any model he prefers, as long as he does so in responsibility for his family, safeguarding their material, emotional and spiritual rights. Violation of these rights would constitute a crime. "Allah", it was revealed to us, "commands justice and equity, kindness and charity to those in relation; and He prohibits adultery, evil and rebelliousness. Would you not mind His commandment?" (60:90). Such is the nature of Islam's position on knowledge. It does not alter the position that Muslims have not observed its principles during their recent decay. Nor do the views count which the Muslim's corrupt intellectual leadership has offered us during the last few centuries.

The two ultimate principles — namely: the integrated social order as opposed to that of continual or perpetual conflict; and the equivalence of reason and revelation which enables us to correct their divergence *in percipi* — occupy a place of paramount importance in our minds. They will become all the more crucial as mankind grows less and less able to afford another world conflagration, or to pay the price of mistakes of global proportion. The past is no longer valid when conflicts could be contained and when their victims were limited to a few hundreds or thousands; when peoples were safely removed from one another so that they remained unaffected by events outside their own boundaries. Soon the world will become a single room in which every cry of sorrow hurts all, and every tragedy affects all. Then Muslims ought to be just as God described them: "We have made you a mean *ummah*, that you may witness unto (provide the example for and judge) humankind" (2:143). At that time also, the judgment of God will be based on what people have earned. "Whoever makes an atom's weight of good works will be reckoned for it; and whoever makes an atom's weight of evil will do likewise" (99:6).

Chapter Eight
Elements for an Islamic Anthropology

Saibo Mohamed Mauroof

Saibo Mohamed Mauroof was born in Sri Lanka in 1944. He studied at the University of Ceylon and the University of Pennsylvania. He obtained his Ph.D. Anthropology from the latter university in 1975. His doctoral dissertation was entitled: "A Religious Cult in Philadelphia." Dr. Mauroof taught at the university of Ceylon, following his graduation. Currently he is Professor of Anthropology in the Department of Sociology-Anthropology of Cheyney State College (Cheyney, Pennsylvania). He has occupied this chair since 1970. He is the author of articles which have appeared in various academic periodicals dealing with anthropological subjects such as *Contributions to Indian Sociology* and *American Anthropologist*.

I. *The Problem: An Historical Perspective*

To speak of an Islamic anthropology is to invoke a judgment; it is also to raise rationally the question of the meaningfulness of the academic kit labelled "Anthropology" to Muslim peoples. It is to ask whether a product of the last few centuries of the Euro-American experience can be useful to Muslims; whether that product can be transplanted into the value-framework and intellectual traditions of peoples who for a much longer period have been through the intellectual, moral, socio-political and cultural experience of Islam.

Anthropologists may certainly assume what Muslims and historians have always recognized, that as doctrine and law, Islam and its existential experience are one and the same for all Muslims, whether Arab, Turkish, Pakistani, Indonesian, Nigerian or Moroccan. They may also assume that anthropology can be treated as a single entity. Despite their apparent oversimplification, both assumptions are necessary to begin this discussion. Hopefully the discussion of the substantive issues that such assumptions generate will lead to more realistic assessments of the true dimensions of the issues and units involved.

The post World War II international context and its impact on the contemporary redefinitions of the anthropological task and purposes, as of other human sciences, may be mentioned here. The venue of the Xth International Congress of Anthropological and Ethnological Sciences (New Delhi, 1978) and its programme are indications of the changing times. The recognition, in this context of a changing world order, of an imperialist European nationalism at the root of anthropology, is the argument from which two broad attempts at such redefinition (Hymes, ed., 1972; Asad, ed., 1973) have arisen. Even more recently there is growing concern for the discussion of "third world" anthropologies in the United States. To attempt to describe an Islamic Anthropology is to raise a related argument.

We may first consider recent news items of instances—organizational scenes and contexts rather than epistemological ones—in which the broad categories of Islam and anthropology have been seen to come together. Anthropology is one of the social sciences included in the programme of the Association for the Promotion of Social Sciences in Arab countries, Turkey and Iran. The recent formation of this group with officers from Kuwait and Egypt was announced in *Current Anthropology* of March, 1978 (p. 210). There are in England and the United States associations of Muslim social scientists with sections for anthropology.[1] The membership of both these associations is international, although many of them work in England or North America.

It may also be noted that in some Muslim countries the development of archaeology is prior to and separate from the development of other branches of anthropology. Thus in Pakistan archaeology continues to be practised with enthusiasm while the formal academic study of social, cultural, linguistic and physical anthropology is relatively neglected. Masry (1975) is an indication of a similar trend in Saudi Arabia.

Such facts increasingly common in recent years point to the importance and ongoing development of academic anthropology in institutions of higher learning in many countries with a predominantly Islamic background. The uppermost question in the mind of concerned Muslim educators is how to fit the Western science of anthropology into Islamic education; how to relate it to the Islamic legacy of

117

learning; how to harmonize its methodology and goals with Islamic objectives.

Notice of the historical development of Indonesian and Turkish anthropology over a longer period of time than in all other nations with largely Muslim populations is available in Koentjaraningrat (1964, 1975); Kansu (1946); Ellen (1976); and Magnarella and Orhar (1976). Particularly in Indonesia, but also to an extent in Turkey, the emphasis seems to have been in social and cultural rather than in physical and archaeological anthropology. It may be said of such developments as in Turkey and Indonesia (as of other Muslim countries for which formal histories are not available) that they are post Euro-American-anthropology events; that while the discipline has become institutionalized, it is in a state of underdevelopment rather than development. Further, the practitioners of the craft in these countries owe their training to U.S. and European institutions. Their anthropology is a part rather than a variant of Euro-American anthropologies. They seem to differ little from Euro-Americans in their work even when dealing with Islam (See e.g., Hanifi, 1974).

The organizational framework of museums, many of which are to be found in such countries, in so far as they have incorporated anthropological and ethnological sciences, is evidence of the diffusion of anthropology in Muslim lands.

The need for further developments in one or all subfields of anthropology in Islamic countries is still to be ascertained. Bohannon and Glazer (1973), reviewing the development of anthropology in Western countries, have suggested that it is in part a consequence of the development of the middle classes and their urbane, perhaps humane, sensitivity and ideology. Will a Westernized or even Western style middle class assume the necessary control over the future transformation of Muslim values thereby creating the possibility of a widespread anthropologism among Muslim peoples? This is more wishful thinking than fact.

The importing of Euro-American anthropology or segments of it into Islamic educational and scientific settings could happen in complete ignorance of traditional Islamic values and academic traditions. Not only has this limited the attractiveness of anthropology to Muslim students and followers of Euro-American learning, but it has also stultified the development of the anthropology of Muslim peoples, even within the narrow range of the culture region often demarcated as the Middle East and North Africa. Thus Robert Fernea and James

Malarkey, reviewing the anthropology of this largely Muslim region (1975: 183), affirm

> "that, with few exceptions, contributions to anthropological literature based on Middle Eastern research have failed to have an important impact upon theoretical concerns in the field of ethnology. Nor has there been an appreciable development of a fruitful dialogue between MENA anthropologists and Orientalists.... In addition, anthropological studies from the MENA have largely failed to attract an audience of scholars beyond those devoted to undertaking such studies themselves..."

I have examined several back volumes (dating back to 1950) of *Islam and the Modern Age* as well as *Islamic Culture* (Hyderabad, India), *Islamic Review* (Woking, England), *Islamic Studies* (Pakistan), and *Al Ittihad* (Indianapolis, U.S.A.), looking for any evidence of the influence of anthropological idea-sets in the scholarly contributions to these journals. My attempt to systematically classify the articles in regard to whether they (1) accept, (2) reject, (3) neglect, or (4) appreciate anthropology turned out to be unnecessary. It was clear that the overwhelming number of contemporary Muslim writers and editors are unaware of or ignore anthropology, although they are quite responsive to several other Euro-American scientific and humanistic disciplines. It was quite clear that the anthropological education of Muslims other than those who are anthropologists and publish for other anthropologists, has remained as woefully inadequate as the Islamicist education of anthropologists studying Muslim societies.

Thus, parochialism of MENA anthropology derives from the Western, and hence alien, epistemological perspective hitherto employed. The almost complete absence in MENA anthropology of the epistemological perspective of the peoples studied, particularly of their own Islamic faith and value-sets, of the Islamic intellectual tradition, is rather conspicuous and vitiates their studies.

The Indian situation, briefly in comparison, is quite different. The anthropology of India is a subject of engrossing interest to all anthropologists. Its development may be seen as a model for other non-Euro-American countries in their efforts to produce a modern behavioural science of the study of the human being. Such an admirable state of affairs is in part attributable to the education that Western and Indian anthropologists have received in Indic notions, values, and philosophical assumptions, and the ways in which such ideas have become incorporated into the epistemological framework of the anthropology of India. I am here thinking of the methodology of "text and context" and "great traditions" studies. Other than the possible exception in the use of the paradigm of cyclical change derived from Ibn Khaldun in some studies of Arab society (e.g., Gellener, 1969; Cole, 1975), the anthropology of the Muslim Middle East and North Africa has been deficient in the consideration of theoretical notions derived from the Quran and the Hadith literature, or from the scholarship of the classical age of Islam in seeking explanations for Muslim behaviour.[2]

Another perhaps more informed alternative for introducing anthropology into the Islamic curriculum is for exchange and dialogue to take place between the "Westernized" and "non-Westernized" sides of the process often represented in modern Muslim countries by different segments of an emergent middle class. The cultural awareness of the "non-Westernized" side of Islamic politics is largely unknown in the West. A notable step in the direction of such development for Muslim anthropological studies is evident in the very recent work of the Moroccan historian Abdullah Laroui (1967; 1976). The first of these works has been seen by critics as employing a cultural anthropological methodology. The second is a detailed examination of Von Grunebaum's methodology and conclusions of his study of Islam (1955a and b; 1971), through which he reached the audience that other anthropologists, according to Fernea and Malarkey, will have to find.

Abdullah Laroui's Arab and Islamic response to Von Grunebaum's brand of anthropology is also an expression of an avowed openness; an invitation to the anthropologist to concern himself with things Islamic:

> "If we have no desire for the fragmentation of research to result in a cultural protectionism where each keeps his patrimony for himself and forbids others to touch it, we must submit to new rules of the *munazarah*" (1976: 45–46).

The *munazarah* is to Laroui the dialectical controversy between Orientalist and Muslim intellectual. Other dialectical oppositions could apparently become part of the exchange. The new rules he sets up are clearly for greater communication not only between Euro-American anthropologists and non-Euro-American scholars but also between diverse types of scholars, of the East and the West, such as, for instance, between historians and anthropologists who are concerned with the study of Muslim peoples. The need for such *munazarahs* expressed by Laroui coincides with the needs outlined by Fernea and Malarkey (*op. cit.*). It is hoped that this contribution will be seen as addressing itself to some segments of that general need.

Laroui accepts the basic academic promise of Western style anthropology, although in some ways different from the Turkish and Indonesian anthropologists. His critique of Von Grunebaum assumes most of the latter's methodological principles and disputes his conclusions. It can therefore have little promise for the Islamic educator concerned with the Islamization of social science as a whole. What is needed, therefore, is a measure far more basic and perhaps radical, i.e., the discovery of a methodology for the study of man in history in the religion and tradition of Islam. Such anthropological methodology would then be used to study Muslim life in its various configurations — ethnic, geographical, linguistic, cultural and historical. Where such methodology agrees with Western anthropology there can be a field of mutual understanding and common interpretation, though for the Muslim reader, justification of the methodology would have to remain a purely Islamic affair, embedded in the Islamic tradition itself. Where Islamic methodology disagrees with Western anthropology, dialogue is necessary; but, as Laroui has pointed out, "only after the rules of the game have been determined". Such "rules" cannot be the result of convention, or of unquestioning loyalty to either tradition, but the product of critical meta-anthropological thought to which the Islamic anthropologist is now inviting his Western colleague. The entrance of the Islamic thinker into the anthropological arena signals therefore a clear need for self-re-examination, and provides the challenge necessary for academic health and intellectual renewal. It is a blessing for the discipline as such.

II. *Elements in the Sources of the Islamic Faith*

Any examination of Islamic literature would immediately reveal that Islam and anthropology are not innately contradictory. The survey I have conducted as initiation of the study of this matter produced nothing to change this judgment. Every society has its own folk anthropology, and the world of Islam is no exception. It derives its methodology from the traditional wisdom which properly belongs to it and casts a mine of information about itself, its neighbours and the world under the perspective of that same wisdom. It may even be useful for the so-called modern "scientific" anthropologists to know this simple and obvious truth. Further, the pre-modern anthropology found in the Islamic sourcebooks is more than a set of folk beliefs and perspectives. It is part of a rational, intellectual tradition of major consequence to a belief system as well as to a set of philosophical notions which I believe are still relevant to many Islamic peoples, and viable in any ideational confrontation with the Islamic tradition.

The Quran is the proper beginning point for any examination of Islam. It clearly requests people to observe the remnants of ancient and powerful kingdoms as signs from Allah of what happens to those who refuse to obey the commandments of Allah (III:137; VI:11; XXII:44–46; XL:21, 82; XLII:10). It would seem that in the wake of the revelation of Al Quran Muslims believed in and followed the suggestions for historical and even pre-historical understanding contained in the verses cited above and in many others, e.g., II:132–136; III:137; VI:11; VII:59–182; XXII:44–46, 78; XXXV:31; XXXVIII:3; XL:21, 78, 82; XLII:13; XLIII:78; XLVI:9–12; XLVII:10; LVII:25–27; XLI:6, 14; LXIV:5–6; LXIX:4–12; LXXI: and LXXXVII:18–19. A universal notion of human history seen as a gradual progression of the human receipt of divine messages, messengers, and communities established on that basis, and the eventual downfall of kingdoms and communities that disbelieved, became a part of Quranic interpretation. Sufficient information is provided by the Quran's interpreters to imply that Muslims did indeed envisage a Quranic paradigm of universal history. Gaps in the information provided by revelation particularly regarding pre-Muhammadan times were filled up with information from other Islamic, Christian and Jewish sources (Mahdi, 1971:136). The notion of Islamic history as beginning with Muhammad (SAAS) and tracing the development and subsequent fall of the Islamic empires is apparently foreign to Muslims. The "history" contained in the Quran and elaborated in some of the subsequent developments of Muslim historiography is, in a sense, a much more anthropological history of man. Tabari, Mas-'udi and al Biruni of the 9th and 10th centuries (A.C.) had dealt with problems of such a history three centuries before Ibn Khaldun. Ibn Khaldun's *'Ibar* — the actual history he wrote — is a critique of many of the "histories" that had come about as a result of such historical and cultural curiosity; and it also contains, in Books II to V, histories of ancient nations. Such historical curiosity of a non-sectarian and universal scope declined in later years (see Inan, 1946; Mahdi, 1971; Rosenthal, 1952; Von Grunebaum, 1971). In the early days of Islam, however, the Quran served as an impetus for historical and cultural scholarship. There is much in the revealed religious injunctions of the Islamic faith that made the investigation of things of the human past a matter of importance to the believer.

In addition to suggesting the need for historical understanding, the Quran has also encouraged a scientific approach to the study of history. The Quran has done this as an integral part of a rational and systematic study of all phenomena which it had advocated and made the epistemological foundation of all knowledge, human and divine.

Muhammad Marmaduke Pickthall, in introducing Surah XXX (Al Rum) points out that in the context of their delivery the verses contained in that chapter constitute a specific prediction about what was then the religio-political future of humankind. Further, "the laws of nature are expounded as the laws of Allah in the physical sphere, and in the moral and political spheres . . ." (1977:422). Of the several verses in that chapter that are interesting to the anthropologist, 21 and 22 are particularly noteworthy. In them the human pairing phenomenon of male and female as well as "the creation of the heavens and the earth *and the difference of . . . languages and colours*" (emphasis mine) are pointed out as signs of Allah, and as things to which thinkers and the knowledgeable ought to pay attention.

In addition to providing cues for the study of the laws of nature in the physical as well in the socio-political, cultural and moral spheres, the Quran also recognized group structures. The Quran clearly contains a conception of humankind. Having posited Adam and Eve as the primal originators of human beings, it addresses itself to the "children of Adam" (e.g., in VII:35, 172). The human species so defined is composed of several kinds of groups separated by variations of belief and disbelief. Other natural factors contributing to differences in the nature of groups are implied in the mention and discussion of such entities as the "home" in XXIV:36, "townships" in XLVI:27, and XLVII:13, "castes" in XXVIII:4, 15, and "nations" (of *"jinn and humankind"*) in VII: 38, 39, 168; X:48; XXXV:24; and XLVI:89, 92. There are also, of course, numerous references to "tribe". XXXIII:6 also recognizes friendship, kinship and common faith as three separate levels of natural bonds of mutual obligation. In these statements and in the reference to processes of group corruption

(XVII:58; XXI:6), there is an obvious implication of a social psychology, of social facts known to the modern anthropologist from the works of Emile Durkheim. Such "facts", however, were apparently well known to Muslim scholars and were highly elaborated by Ibn Khaldun, whose *Muqaddimah* is now "counted as the first major work in cultural anthropology" (Honigman, 1976:49).

Euro-American scholars (anthropological historians not excepted) have not paid sufficient attention to the relevance of Quranic and post-Quranic Islamic idea-sets in their evaluation of Ibn Khaldun's genius. Rosenthal (1958:xxxvi) even suggests that, while Ibn Khaldun's body may have been Arab and Muslim, his mind was Spanish! The idea that he is a "lone star shining brightly in the darkness of the medieval world" is consonant with European notions of world history and the development of science and technology in the world. Such notions are understandably biased in favour of European superiority. Muhsin Mahdi (1971), being more aware of the history of Islamic thought, is able to place Ibn Khaldun's work within the framework of the development of Islamic historiography, religious thought and philosophy. He has analysed (1971:133–170) the works of Tabari (d. 923) and Mas'udi (d. 956) of the fourth century after the Hijrah, of Miskawayh (d. 1030), al Biruni (d. 1038), Sa'id (d. 1070) of the 5th century A.H., and Ibn al Athir (d. 1233) of the 7th century A.H., all of whom preceded Ibn Khaldun, as well as the work of Khafaji and Sakhawi of the 9th and 10th centuries A.H., who wrote after Ibn Khaldun.

Before we go on to review some of these works in further detail, a relevant note on the Hadith should be made. After the Quran, the Hadith – the collection of sayings attributed to the Messenger of Islam (SAAS) – is the second most significant source of Islamic knowledge and wisdom. The method of transmission of this body of information is oral to a much more significant degree than the Quran, which became an agreed and written text very early in the history of Islam.

Within the first couple of centuries after Muhammad (SAAS), however, the authentication of the sayings as in fact emanating from him became a serious issue for Muslim scholars. Elaborate methodologies for ascertaining the veracity of individual sayings, the context of their delivery, and their implications for generalized rules of philosophy and knowledge were developed. In consequence, textual criticism, textual history, and biography were born as critical sciences with elaborate and sophisticated methodologies. Detailed documentations of several aspects of the process are available (Rahman 1968:43–74). Historians of ethnographic methodology can thus look to the comparative studies of developments in the "Science of the Hadith" for perspectives in the modern experience of ethnographic data collection and ethnological theory-making, both of which are based on orally transmitted cultural information.

III. *Elements in the Sourceworks of Islamic Culture and Thought*

Further, Muslim "social scientists" prior to, contemporaneous with, and succeeding Ibn Khaldun projected a universalist historical awareness, certainly under the Quranic injunctions noted above. Al-Faruqi (1977), among others, has noted that such awareness was also of peoples other than the Muslims and their rulers of that time. Information pertaining to strange and foreign peoples based on "eyewitness accounts" became an integral part of some of the studies of society, culture and history. The availability and use of such proto-ethnographic information at that time is significant for an Islamic anthropology and especially for the place of ethnography in it.

Several areas in the ideological consciousness of the Muslims before the time of Ibn Khaldun, such as the search for reliable tools for the investigation of the oral history of nascent post-Hijrah Islam, the methodology of abstracting generalized rules of behaviour from the wisdom contained in that oral history (and in the Quran) and the need to know more concerning ancient history primarily in regard to pre-Hijrah Islam, provided a framework for investigations, the scope of which was not very different from the scope of certain phases of modern Euro-American anthropology. These intellectual problem areas became even more apparent in the socio-political context of an empire that trade and conquest had placed face to face with the various divergent customs, languages and cultures of an extremely large portion of the human species.

Currents of information generated by such complexes had produced, even before the *Muqaddimah*, important works on human geography,[3] and even more general contributions toward the understanding of the relationship between character, culture, and physical environment. As Mahdi (*op. cit.*:143) has shown, the influence of Greek ideas is also important for understanding Masudi's shift in the emphasis of Islamic historiography from political chronology to world cultural history. Masudi was specifically concerned with the relation of natural environmental factors to human historical developments, a theme that Ibn Khaldun was to elaborate later. Masudi was also concerned with parallels between cycles of plant and animal life and human institutions, although only at the analogic level.

Even more relevant to the understanding of the parallels between Islamic scholarship prior to Ibn Khaldun and modern anthropology, is the work of Said (d. 1070). His *Classes of Nations* deals with the "history, learning, character, and social life of various nations."[4] In that work, he attempted to develop a general anthropology relating diverse human institutions to various faculties of man. He anticipated by centuries the 19th century (A.C.) evolutionary anthropologists in his method of classification of "civilized" and "barbarous" peoples on the basis of the cultivation of the sciences as expressions of human rationality.

Said lived in Spain about the same time as al Biruni lived and wrote the classic *Indica* in what was then the eastern outpost of the Islamic empire. Both of these scholars concerned themselves with the rationality of other, non-Islamic peoples. The continuing relevance of al Biruni's work to Indic studies is shown in the recent reissue of Sachau's translation of the *Indica*, which was edited and introduced by professor Ainslee T. Embree (1971). Some features of al Biruni's anthropology may be briefly noted. He emphasized the "ideational" dimension of culture, but not to the point of excluding the "phenomenal", to use Goodenough's (1964:1–24) separation of levels of cultural reality. His purpose was to write a book on "another people", namely, the Indians, for his fellow Muslims – a vision that resembles rather closely that of the anthropologists. In doing so, he has shown himself to be remarkably free of ethnocentric bias. Indeed, al Biruni's work is a model of the phenomenological school of comparative studies, fulfilling to near perfection the school's demands for *épôché*, for respect for and empathy with the material studied, for recognition of its *sui generis* character, for a real and adequate intuition of its essence and determinative structure.

Al Biruni's work on the India of his time (the 11th century A.C.) is vast and contains copious quotations from Sanskrit texts – a tendency of even the present day anthropology of India—in addition to other behavioural information on many subjects including Indian philosophy, morals, astronomical and mathematical sciences, mythology, and astrology. In presenting the results of his participant observation of North Indian society, he often prefaces his discussion with philosophical and comparative comments drawn from his knowledge of ancient Greek, Christian, Persian, Roman and other sources. In dealing with the political, social and religious institutions of Indian peoples, he often compares them to contemporary Islamic institutions and points out variations of belief and practice among Hindu peoples.

Thus, in describing religious belief, he makes a distinction between the beliefs of the "educated" and "uneducated" people (Sachau I:27, 111). In spite of the caste system, al Biruni emphasizes the Hindu philosophers' conviction of the essential equality of all human beings (II:137, 138). In regard to religious belief the common people were not only separable from the elite, but exhibited a great deal of variety among themselves. Some of their ideas were abominable, "but similar errors also occur in other religions . . ."; and Islam, he warns his reader, is no exception (I:31). In examining the religio-philosophical notions of the educated, he is quick to see the similarity of some Vedantic notions to the ideas of the ancient Greeks and of the Muslim Sufis (I:33ff). In contrasting Islam with Hinduism, he sees the Islamic *shahadah*, the statement of belief, to be the proper point of contrast in the understanding of the Hindu doctrine of reincarnation (I:50ff). He analyzes principles of marriage, for example, comparing polyandry in Hindu mythology with the marriage customs of Arabs and ancient Iranians (I:108ff; II:154ff). His book contains a theory of the origin of castes in India, seeing it as a product of the combination of religion and state, and compares Indic caste with the caste system of ancient Iran (I:99ff; see also II:130ff).

The above mentioned are not the only scholars who may be seen as predecessors to an ethnographic component of an Islamic anthropology. There are others whose works are available in translation and gathering dust in Euro-American libraries. Among them is the work of Ibn Hawqal, a 10th century (A.C.) Arab traveller in Persia whose work on "Persons (also of the Manners, Languages, Religions, and Chief Families) of the People of Fars" was translated by Sir William Ousley in 1800. There are also works which have not been translated or are only recently being translated. The value of such translations for unravelling the history of Africa (and of African social conditions) prior to European penetrations, is now beginning to be felt (Davidson, 1970:ix; Koch, 1977).

Among the sources that Koch draws on for reconstructing the history of Timbuctu is Ibn Batutah, a 14th century (A.C.) Moroccan-born world traveller who wrote detailed accounts of his observations during travels in several regions of Africa and Asia (Gibb, 1929; 1958; 1962). Ibn Batutah recorded impressions of history, politics, religion and other topics of ethnographic interest regarding Mosul (Gibb, 1929:103), Bursa (136), Istanbul (160), and Multan (188–189), among other cities of the world in the 14th century. The account of his travels include descriptions of 14th century West African cities and customs, Muslim and non-Muslim (318–328), as well as references to several ancient African cities (e.g., p. 58). His narrative also contains descriptions of Chinese places, customs and peoples (283, 297). In Delhi, Ibn Batutah served in the court of the sultan whose gate was "never without some poor man enriched or some living man executed" (197). Combining the role of judge, courtier, world traveller and diplomat, he traversed the entire breadth of the Islamic Empire of his time; and his account is foremost among the descriptive accounts of Muslim society in the second quarter of the 14th century A.C.[5] He was commissioned to write the account of his travels at the end of his journeys by the royal court at Fez.

Ibn Khaldun also served in that court briefly and seems to have been rather sceptical about Ibn Batutah's tales. He hardly mentions the famous traveller in his work (see Rosenthal, 1958:369–372). Perhaps even the keen sociological and cultural acumen of Ibn Khaldun was not imaginative enough to grasp the Oriental wonders of what was to him an *other* world. Ibn Batutah, however, was quite at home in India and China as much as he was in the Western realms of the Islamic commonwealth of that time. Much like European travellers of later centuries whose accounts have laid an important cornerstone for the rise of ethnographic theory and method, he had an eye for the strange, exotic and unusual.

Being himself a firm believer in the extra sensory perceptions and miraculous powers of the saintly (see Gibb:36ff), he faithfully records experiences with Indic yogis (226) and Chinese magicians. He mentions Muslims who ate hashish and thought "no harm of it" (124) and relates incidents of violence including suicide (124,191).

Like other travellers' accounts, Muslim as well as European, Ibn Batutah's is full of structural details and bits of information useful for reconstructing ethno-histories of the regions and peoples he visited. Although one may look in vain in his writings for a general theory of human behaviour, these writings leave no room for doubt that Ibn Batutah was keenly aware of the basic premises of Islamic anthropology enunciated in the Quran and the Hadith, or rationally established by the jurists as bases of Islamic laws. Ibn Batutah, like Ibn Khaldun, was a jurist by occupation. Issues in *fiqh* and jurisprudence frequently come up in his descriptions.

The ethnographic content of his descriptions, however, are not diminished, but added to, by his concern with Islamic legal traditions — as, for example, in describing the manner of jumah prayers in Jeddah (106), or funeral processions in Damascus (71). He is quite tolerant of Shii customs (82), of joint Muslim, Christian and Jewish processions (68, 69), and describes in fair detail Christian monasteries (162), making special mention of a king who became a monk (163). He is critical of certain deviations from the religious law in the Maldive Islands (24ff) and in West Africa (318–328).

He is also critical of Indian Muslim monarchs and their atrocities and recounts the story of his intervention on behalf of an "infidel" who was to be "cut in two" (see pp. 224, 263). He provides much information on ocean trade contacts linking East Africa, Arabia, the Persian Gulf Area, India and the further East (106ff), and on popular beliefs regarding venerated sites celebrating heroes of pre-Hijrah Islam (55, 57–59, 62, 63).

It seems clear that Ibn Batutah, the sponsors of his publication, as well as his readers of the 14th century, display an attitude of amazement and wonder at the picture of diversity of shapes, colours, sizes, languages, habits and customs of the human being. The information that Ibn Batutah has recorded documents not only variations between "believer" and "infidel", but also variations within the world-wide Islamic community of his times. His observations on behavioural complexes following from Muslim stratification systems are particularly interesting. They include information on slavery (30–31, 70, 85), "negro" and "white" relations in India (101), women in relation to men in Turkey (146), and women and slaves in relation to men (136, 171, 196). Being much travelled in East and West Africa, he carefully lists the "many admirable" and the "bad" qualities of black Africans.

All of this may point to an attitude of tolerance toward sociocultural diversity. Such attitudes, which could certainly be derived from the fundamentals of the Islamic system of faith and ethics, were needed to maintain the poly-ethnic composition of the world-wide Islamic community and were perhaps fortified by the institution of the *hajj* (pilgrimage). Later, European world-wide dominance and fascination with non-European peoples was to produce, for the West, part of the framework necessary for a science of anthropology, which Islam and its thinkers had provided a millennium or more ago. In the Islamic context, Muslim scholars studied their own as well as other societies, adding to human self-knowledge and to the disciplined search for the same.[6]

We may also ask whether the fact of socio-cultural diversity, so apparent to Muslim scholars, ever posed a problem of a philosophical or political and practical nature to Muslim societies and their scholars. The Quran had taught them that Allah—*subhanahu wa ta 'ala*—has willed the creation of humankind as tribes and peoples, different in endowment and potential, that humans may know, complement and cooperate with one another (Quran 49:13; 4:1). The Prophet (SAAS) had specifically declared that all humans proceed from Adam, and Adam from dust; that no Arab is superior to non-Arab, and no white to black, unless it be in righteousness. In other words, Islam recognizes, first, that ethnic diversity is a positive good; second, that it constitutes no basis for discrimination among humans; and third, that righteousness—rather than ethnic differentiation—is the true basis for merit among humans. The question therefore becomes whether or not the Muslims did differentiate between one another ethnically in their daily intercourse. Here, it must be observed that the *shariah* or law of Islam regulated those deeds which affected one another in their lives, properties, wealth or labour. Deriving directly from the Quran and *sunnah* (practice of the Prophet) the *shariah*, as it has come down to us, is absolutely free of ethnic discrimination. Moreover, being overwhelmingly comprehensive and pertinent to every aspect of human activity, the greatest portion of human life under Islam was free of discrimination. This is the serious portion. Its real consequences can be observed in any Muslim city today where most or all Muslim ethnic types parade on the streets in total unconsciousness of one another's ethnic differences. Evidently intermarriage, colour-blindness, and ethnicity-unawareness have combined under Islamic law to create the present universalist configuration. Practically every Muslim city has seen countless waves of immigration by ethnically diverse peoples; but none is known to have had an ethnic ghetto for any length of time, and none has one today. Certainly Islamic religion, culture and law were the dissolving agents of ethnic differences.

There remain the realm of academic speculation as to how the ethnic groups came to be what they are, and the realm of *belles lettres* and recreation where a poet or social analyst gives vent to humour or personal like or dislike, and makes poignant psychological observations characterizing one or another person or group. One could imagine such remarks to be met with smiles or flaring tempers; but their effect would in any case be shortlived when seriousness resumes. It is on this level that most discussions of ethnic diversity in the world of Islam occur; and it is here where orientalists blunder by falsely extrapolating from the realm of *belles lettres* to that of living practice. An Anglo-American Orientalist has tried hard recently to establish the existence of racial or colour discrimination in the Muslim East (Lewis, 1971). His attack on the notion that Islam is non-discriminatory in norm and conduct seems to have been prompted as much by the rise of Black nationalism in the West as by the traditional antagonism of orientalists and missionaries. To these Lewis adds the opposition of Zionism. Lewis has compiled an impressive list of literary allusions to "black" and "white" as "evidence of prejudice and discrimination in Islamic countries," thus committing a *non-sequitur* at once unintelligent and misleading.

Ibn Khaldun devoted two sections of his *Muqaddimah* to the discussion of colour as a factor of importance in natural human variation (Rosenthal, 1958:I, 167–173). Speculating on the origins of ethnic diversity, he related skin colour to temperament, intelligence, etc., basing his view on a theory that we might today call ecological. Refuting the Biblical notion that Negroes are black because they are the descendants of an "accursed Ham", he puts forward the notion that both black and white people who inhabit regions far removed from the "temperate zones" have a different character from that of the rest of humanity. Interestingly, to him it was the air that people breathed, rather than the sunlight that they were exposed to that affected the colour of human skin. In continuation of al-Idrisi's thinking, Ibn Khaldun divided the world into various zones by climate which corresponded to human physical and psychological traits, thereby linking with and developing notions inherited from Greek antiquity. These notions not only relate to culture as a normative system, but provide an explanation for how human variation comes about, persists and changes.

IV. *Towards an Islamic Anthropology*

The sources that we have drawn upon to make up an outline of a proto-anthropological strand in classical Islamic knowledge are primarily religious; they are also from a literature classified by Islamicists as *adab*, which is close to what we might call "humanistic studies" with a literary base (al Faruqi, 1977). Studies pertaining to the descriptive, analytic or comparative investigation of man have never been a separate subject in the Islamic curriculum. There is therefore no tradition of an Islamic anthropology as such. Islamic anthropological knowledge is derived from materials produced in the service of disciplines other than the study of man. Ibn Khaldun's effort to create such a science was made while he himself served as a professor of law and jurisprudence. With him, as well as with other luminaries of traditional scholarship in legal and jurisprudential studies, variations in application of the *shari'ah*, or Islamic law, served as a major emphasis in the study of what we now call human variation. Besides the Quran and the *Hadith*, the repositories of normative Islam, the vast legacy of law and jurisprudence must be studied if we are to achieve an understanding of the practical and official Islamic point of view on human variation. The *adab* literature, literary and speculative, which often figured prominently in the discussion of orientalist and Muslim philosophical issues (Mahdi, *op. cit.*, 133ff), is entertaining but not instructive about the Islamic point of view.

Here is an interesting problem for students of Islamic culture of the past and the present. The future development of the Islamic study of variation needs a proper elucidation of the anthropological contents of all three levels or kinds of literatures: the scriptural, the juristic and the *adab*. The first is normative, the second is descriptive in most areas of human activity; the third is mostly speculative. This content needs to be enlarged by the combination of the perspectives of modern studies of antiquity and human biology. Such enterprise requires basic attitudinal and ultimately philosophical or religious changes in the persons and policies that support such research and study.

The recent introduction of anthropology into Islamic colleges seems to have taken a less revolutionary and more comfortable stance in two ways: the first is to coalesce with sociology as suggested, e.g., in Koentjaraningrat's discussion of the Indonesian position (1964). The second is for anthropologists to engage in studies of so-called "tribal" peoples (e.g., Akbar Ahmad, 1977). On another level we may note the simplistic view of a University of Bombay anthropologist who, writing in *Islam and the Modern Age*, has suggested that all Muslims ought to become cultural relativists and adopt the stance of the ethnographer so that they could overcome their cultural prejudices and thereby promote social harmony in the context of the plural society of the Indian republic.

None of this will do; and the Muslim anthropologist stands under the terrible charge of prostituting his mind and betraying his tradition if he continues to ignore the Islamic anthropological legacy and to think, write and teach as if he were a Westerner. Whatever the outcome of his confrontation it must take place if he is to gain respect and be given serious consideration at all. But no confrontation *can* take place in ignorance. Hence, the task of re-presenting the legacy of Islamic thought systematically, under categories consistent with modern discipline, for the ready use of reason and understanding. This should be done in analytical articles, anthologies of readings, as well as in book-length studies of each of the dominant ideas characteristic of the system of Islamic thought. Only after all this has been done well, i.e., it has built up such a body of literature that there is no room to suspect that any facet has been neglected or poorly understood, that the bigger and harder task of critical evaluation of the Islamic and Western legacies of anthropological thought is to begin. Naturally, there is no telling *a priori* how the critical evaluation will proceed or where it will lead. The rules of *munazarah*, as Laroui had said, will have to be elaborated and established, and the main outlines of a metaphysic of anthropology drawn out.

In this work, the Muslim anthropologist in quest of an Islamic anthropology, would have to have recourse to the Quran, therefrom to draw his main categories, both theoretical and axiological. Anthropological studies referring to Quranic prehistorical allusions would seem to be a sure and certain area in which modern research can find a listening and responsive ear from Islamic quarters. For example, the meaning of Iram—a southern Arabian site specifically recommended for visiting—has been briefly discussed on the basis of scientific as well as religious theories by commentators (Pickthall, 1977 fn. 2, Surah LXXXIX). Considering the number of verses available in the Quran which tend to encourage archaeological and prehistoric investigation, it is surprising that such enterprises have not received wider recognition in the Islamic religious scholarship of recent times. The need for Muslims as nations, peoples and intellectual communities to engage in the study of antiquity for reasons of piety and religion if for no other reason has been strongly pointed out by Professor Isma'il Raji Al-Faruqi (1962). Dr. Al-Faruqi's paper is decidedly pro-Islamic; anthropologists may not necessarily agree with his assumptions regarding the interrelationships of beliefs, prophets and socio-historical entities. However, his is a contemporary appeal sensitized by and reacting to a body of investigation and interpretation that forms a crucial part of modern anthropological knowledge. In making such an appeal, Al-Faruqi reopens discussion of an old problem of Muslim historiography. He envisages a prehistory of Islam, or a "pre-Hijrah Islam", a notion of Islamic or proto-Islamic communities prior to Muhammad (SAAS); of a period of Islamic history that is unwritten in the sense that anthropologists use that expression, hence available mainly through archaeology. Although Al-Faruqi does not mention prehistory or anthropology by name, there is in his argument a clear acceptance of several basic modern anthropological notions of time and of the human phenomenon. To the extent that Muslims of the present day world have a need to rationalize or, better still, to understand at a deeper level their belief sets in the areas of history and prehistory, they have an urgent need to understand, even if it is for purposes of refutation, the theoretical and methodological premises of modern archaeology and anthropology. Dr. Al-Faruqi clearly sees this.

A second area in which anthropological study receives Quranic backing and encouragement is biological anthropology. That man has developed religiously and culturally, and in his adaptation to nature and environment, is a necessary premise of the Quranic theory of revelation and prophecy, a postulate of the history of human righteousness. The Quran hardly presented righteousness in terms of rites and rituals but in those of real action, the domain of interaction with nature, whether plant, animal or human. For example, the biological principles that are a segment of the modern anthropological explanation of the non-human part of human history, it would seem, are quite within the Islamic framework, as a Pakistani student of the Quran has recently concluded from his study. Attempts to juxtapose biological theory with the Quranic theory of life have also been made by Khan (1976; 1977). These are hardly known statements, despite their high relevance for Islamic anthropology, as are the statements derived from the Quran itself over several centuries of Islamic rationality. However, they are available as instances of a process that must become very much part of the future discussions of Islam, of the emerging Islamic discipline of anthropology, and of its significance for Islamic education.

A clue to understanding the religious faith of the contemporary Muslim is the essentially anthropological basis of the problem that al-Faruqi poses (*op. cit.*) regarding the "Islamicity" of Akkad and Ikhnaton; and of the question that Mahmud Muftic poses in asking (1971) whether the Prophet Idris mentioned in al Quran (XIX:56, 57; XXI:85) is the same person as Imhotep of ancient Egypt.[7] Such questions rest on the acceptance of pieces of modern knowledge that is as much a part of anthropology as they are of Near Eastern history. More to the point is the assumption of a kind of universal Islamic "psychic unity of mankind", a basic modern anthropological paradigm.

Philosophically the issues that Al-Faruqi raises are crucial to the discussion of human evolution. As Wadud (*op. cit.*) would have it, there are Quranic statements which may be interpreted as verifying the notion of evolutionary processes. As we have suggested earlier in this paper, the Quran certainly presents itself as occasioned by a long period of gradual religious and moral evolution of humankind. Pickthall (1975:v) briefly notes the manner in which the Quran states the proposition.

In some brands of Sufi thought, devoted to the psychology of individual salvation and spiritual ascendency, such evolutionism has been seen as pertaining most specifically to religious and moral elevation in the human — as stages of universal human historical evolution, as well as stages in the individual development of single humans. Even in such individualistic Sufi notions, there is an analogical and metaphorical similarity seen between stages of individual and group liberation which correspond to bodies of doctrine, the sum of which includes all the religious doctrines of man.[8] (For related notions in Ibn Khaldun, see Rosenthal, tr., I:184–245; III:76–103.) Such notions of psychological evolution are, we may note, an entirely separate segment from his more well known idea of the transformation from "bedouin" to "civilized" forms of existence.[9]

Western anthropologists, however, have manifestly ignored the Islamic legacy of anthropological thought except in the very few and circumstantial instances noted above. This is perhaps due to an early bias in the Christian understanding of Islam. While the Quran and Muhammad (SAAS) proclaimed the message of Islam as a final crystallization of a long series of progressive divine messages, including the message communicated by Jesus (*'alayhi al salam*). The Christian Church has for long seen the Prophet Muhammad (SAAS) as a heretic deviating from the teaching of Jesus. Further, Christian critics have misunderstood the universalism of Islamic teachings contained in the repeated allusion to pre-Muhammadan prophets and mystics and their teachings as in fact evidence of a lack of originality. Notions derived from such theological criticisms of Islam are still discernible in the "objective" studies of Islam in cultural anthropology.

It is not at all my intention to try to claim that the anthropological thought that the Quran presents is the same as or similar to the picture painted by Western anthropology, or by some schools of cultural anthropology (cf. Wadud, *op. cit.*). It is interesting, however, that the Islamic revivalism of the nineteenth century never saw itself or Islam running counter to the investigations of science or development of its engineering of nature. It is also true that the World of Islam has known no parallel to the intense *religion vs. science* debates so characteristic of Euro-America. Perhaps the metaphors of creativity as expressed in the Quran pose a set of problems to the rational mind so entirely different from the ones posed by the Bible that the creation vs. evolution argument, for example, would seem artificial.

The problem of Islam's past contributions and modern world-presence as a continuing and viable ideological option (e.g., Wolf, 1951; Aswad, 1970; Diener and Robkin, 1978; Gulick, 1976) cannot be separated from Western deep-rooted ideological biases and unconscious prejudices which are immediately apparent to Muslims. The identification of Christian ideologies as well as evolutionary and anthropological theorization with one segment of humanity, the Euro-American kind, and the corresponding identification of Islam with another kind of notion of humankind and its evolution, pose a basic contradiction with which the conscientious Muslim anthropologist cannot live. The contemporary growth of several developmental and progressivist ideologies based on diverse indigenous, national and international social psychologies in the modern Muslim World makes the conflict even more vivid. Within the context of this confusion, the intellectual task of unravelling what the Islamic position is, can hardly be said to have begun.

NOTES

1. The U.S. Association is the group that initiated the invitation which led to the writing of this paper. In 1978, the Association held a special Sociology-Anthropology Group meeting in which Dr. Arif Hasan, a physical anthropologist at Indiana State University, read a paper on "Ecological Approach in Medical Anthropology and Its Relevance to Islam" (1978).

2. An important exception is the work of Gustave von Grunebaum (e.g., 1955a and b; 1971). The body of his work, however, relates more to the historical date pertaining to Muslim societies and cultures rather than the ethnographic. Further, perhaps due to as many non-academic as academic reasons, von Grunebaum's initial attempts to develop a series of studies of Islamic civilization in anthropology parallel to the study of Indic civilization have been much slower to germinate and multiply than the efforts of anthropologists who worked on India.

3. The only mention that Harris (1968) has made of the contributions of Islamic scholarship toward the development of anthropological ideas is to refer to the world geography of Idrisi (12th century A.C.). Nafis Ahmad (1972, first pub. 1947) is a useful survey of other geographical contributions. The more recent Tibbets (1971) on the related field of navigation is very valuable.

4. Mahdi, *op. cit.* See p. 144, n. 1, of the 1935 French translation.

5. I understand that Ibn Batutah's work is now included in the history curriculum of high school level education in many eastern and western African countries.

6. There is even a notice of a 19th century Arab study of European music (Cachia, 1973).

7. "Idris" is also part of Ibn Khaldun's discussion of antiquity. See Rosenthal, 1958:II, 365. On a search similar to that of Muftic, see Rofe (1956).

8. In Mauroof (1976), I have analyzed the teachings of a contemporary "Sufi" presenting such a framework.

9. See also Bosch (1950).

REFERENCES:

Al-ittihad, Indianapolis, U.S.A.
Anthropology Newsletter, Washington, D.C., U.S.A.
Current Anthropology, Chicago, U.S.A.
Islam and the Modern Age, Delhi, India.
Islamic Culture, Hyderabad, India.
Islamic Review & Arab Affairs, London, England.
Islamic Studies, Lahore, Pakistan.

Ahmed, Akbar S., 1977, *Social and economic change in the Tribal Areas, 1972–1976*. Foreword by Nasirullah Khan Sabar, Karachi, Oxford U.P.
Ahmad, Nafis., 1972, *Muslim Contributions to Geography*. Lahore, Pakistan.
Al-Faruqi, Isma'il, R., 1962, Towards a historiography of Pre-Hijrah Islam *Islamic Studies* I (ii): 65–87.
– – 1977, Islam and the Social Sciences in *Al-ittihad*, 14 (1–2): 38–40.

Asad, Talal, Editor, 1973, *Anthropology and the Colonial Encounter*. Ithaca Press, London. Reprinted 1975.
Aswad, Barbara C., 1970, Social and ecological Aspects in the Formation of Islam. In *Peoples and Cultures of the Middle-East* vol. 1. Ed. L. E. Sweet: 52–73. Garden City: Natural History Press.
Bezirgan, N. A. 1974, The Islamic World. pp. 375–387 in T. F. Glick ed. *The Comparative Reception of Darwinism*: 375–387, Austin: University of Texas Press.
Bohannon, Paul and Mark Glazer., 1973, "Introduction" in *High Points in Anthropology* ed. with an Introduction by P. J. Bohannon and M. Glazer. New York.
Bosch, Kheirallah, 1950, Ibn Khaldun on Evolution. *The Islamic Review and Arab Affairs*, XXXVIII (5): 26.
Cachia, Pierre, 1973, a 19th Century Arab's Observations on European Music (Faris al'shidyaq), *Ethnomusicology*, 17(1): 41–51.
Cole, Donald P., 1975, *Nomads of the Nomads: The Al Murrah Bedouin of the Empty Quarter*. Chicago, Aldine.
Davidson, Basil, 1970, *The Lost Cities of Africa*. Revised Edition, Boston.
Diener, Paul & Eugene E. Robkin, 1978, Ecology, Evolution, and the Search for Cultural Origins: The Question of Islamic Pig Prohibition with CA comment, *Current Anthropology*, 19(3): 493–540.
Ellen, R. F., 1976, The development of anthropology and colonial policy in the Netherlands 1800–1960, *Journal of the History of the Behavioral Sciences*, 12(4): 303–324.
Embree, Ainslie T. ed., 1978. *Al-Biruni's India*, Translated by Edw. C. Sachau. Abridged, edited, with Introd. and Notes by A. I. Embree. New York.
Fernea, Robert A. and James M. Malarkey, 1975, Anthropology of the Middle East & North Africa: A Critical Assessment, in Siegel, Bernard J., Alan Beals, & Stephen Tyler ed. *Annual Review of Anthropology*, 4: 183–206. California.
Gellner, E., 1969, *Saints of the Atlas*, London.
Gibb, H. A. R., 1929, *Ibn Batuta, Travels in Asia and Africa 1325–1354*, London.
– –, 1958, *The Travels of Ibn Batuta A.D. 1325–1354*. Vol. I, Works Issued by the Hakluyt Society, Second Series, No. CX. Translated with revision and notes from the Arabic text, edited by C. Defremery & B. R. Sanguine. Cambridge Univ. Press, England.
– –, 1962, *ibid*. Second Series, No. CXVII.
Goodenough, Ward H., 1964, Introduction, in *Explorations in Cultural Anthropology: Essays in honor of George P. Murdock*, ed. by W. H. Goodenough: 1–24. McGraw Hill.
Gulick, John, 1976, *The Middle-East: An Anthropological Perspective*: 163–174. California.
Hanifi, Jamil, 1974, *Islam and the Transformation of Culture*, New York.
Harris, Marvin, 1968, *The Rise of Anthropological Theory. A History of Theories of Culture*. New York.
Honigman, John Joseph, 1976, *The Development of Anthropological Ideas*. Dorsey Press.
Hymes, Dell, ed., *Reinventing Anthropology*. New York.
Inan, Muhammad 'Abd Allah, 1946, *Ibn Khaldun: His Life and Work*, Second Edition, Lahore.
Kansu, S. A., 1946, The place of anthropology and ethnology in Turkish universities, and works and studies carried on in that field, *Man*, 56(115): 141–142.
Khan, Majid Ali, 1976, Nature of Life in Biology and in Islam, *Islam and the Modern Age*, VII (2): 42–63.
– –, 1977, Science and Islam on the Origin of Life, *Islam and the Modern Age*, VII (1): 63–85.
Koch, Klaus-Friedrich, 1977, Fabulous Timbuktu, *Natural History*, LXXXVI (5): 68–75.
Koentjaraningrat, 1964, Anthropology and Non-Euro-American Anthropologists. The

situation in Indonesia, in W. H. Goodenough *Explorations in Cultural Anthropology* etc.: 293–323. McGraw Hill.

— —, 1975, Anthropology in Indonesia: A Bibliographical Review *Koninklijk Instituut voor Taal-Landen Volkenkund Bibliographical Series*: 8, Gravenhage, Martinus Nijohoff.

Laroui, Abdullah, 1967, *L'ideologie arabe contemporaine*. With a preface by Maxime Rodinson. Paris.

— —, 1976, The Arabs and Cultural Anthropology: Notes on the Method of Gustave Von Grunebaum, in *The Crisis of the Arab Intellectual; Traditionalism or Historicism* 44–80. Translated from the French by Diarnid Cammell. Univ. of California Press.

Lewis, Bernard, 1971, *Race and Color in Islam*. Illustrated. Harper Torch Books.

Magnarella, P. J., and Orhar Turkdogan, 1976, The development of Turkish Social Anthropology. *Current Anthropology*, 17(2): 263–274.

Mahdi, Muhsin, 1971, *Ibn Khaldun's Philosophy of History: A Study in the Philosophic Foundation of the Science of Culture*. Phoenix Edition, Univ. of Chicago Press.

Masry, Dr. Abdullah H., 1975, Introduction in *An Introduction to Saudi Arabian Antiquities*: 13–16. Dept. of Antiquities and Museums, Ministry of Education, Kingdom of Saudi Arabia, 1395 A. H., 1975 A.D.

Mauroof, Mohamed, 1976, *The Culture and Experience of Luminous and Liminal Komunesam*. Ph.D. dissertation, Univ. of Penna. Department of Anthropology, Philadelphia.

Muftic, Mahmud, 1971, The Prophet Idris of the Quran is the Same as Imhotep (3000 B.C.), the builder of the enormous Step Pyramid at Sakkara, Egypt. *Islamic Review and Arab Affairs*, Jan. 1971: 16–19. London.

Ousley, Sir William, 1800, Translator. *The Oriental Geography of (al-Istakhri) Ebn Haukal, an Arabian Traveller of the 10th Century*. London.

Pickthall, Mohammed M., 1975, *Holy Quran, English Translation*. Karachi, Pakistan.

— —, 1977, *The Meaning of the Glorious Qur'an. Text and Explanatory Translation*. Muslim World League – Rabita, Mecca Al-Mukarramah, Saudi Arabia.

Rahman, Fazlur, 1968, *Islam*, Doubleday Anchor Edition, Garden City.

Rofe, Hussein, 1955, "Lord of Horns": A Quranic Prophet. *Islamic Literature*, VII (11 & 12). Lahore.

— —, 1956, *ibid.*, VII(1).

Rosenthal, Franz, 1952, *A History of Muslim Historiography*, Leiden.

— —, 1958, *Ibn Khaldun: The Muqaddimah; An Introduction to History*. 3 volumes. Bollingen Series. XLIII.

Sachau, Edward, 1964, *Alberuni's Indica*. With Notes, Introduction, and Index. Bombay, India. (First Indian Edition, 1865).

Tibbets, G. R., 1971, *Arab Navigation in the Indian Ocean Before the Coming of the Portuguese*, etc. Royal Asiatic Society of Great Britain & Ireland, London.

Voget, Fred W., 1975, *A History of Ethnology*, New York.

Von Grunebaum, G. E., 1955a, *Islam: Essays in the Nature and Growth of a Cultural Tradition*. London. (Also published in a special issue of *The American Anthropologist*, LVII, 1955).

— —, 1955b, Ed. *Unity and Variety in Muslim Civilization*, Chicago.

— —, 1971, *Medieval Islam – A Study in Cultural Orientation*. Second Edition. The University of Chicago Press.

Wadud, Sayed Abdul, 1971, *Phenomena of Nature and the Quran*, Lahore, West Pakistan.

Wolf, Eric, 1951, The Social Organization of Mecca and the Origins of Islam. *Southwestern Journal of Anthropology*, (7:101–124).

Part 2 Natural Sciences
Edited by Dr Abdullah Omar Nasseef

Preface

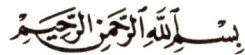

In the 20th century the word "scientific" is regarded as if it were a magic wand. Anything classified as "scientific" is considered to be true, honest and free of bias, but anything "unscientific" is looked upon as suspect. This illustrates the great respect that is paid to the scientific approach to knowledge — or methodology — which scientists claim to have evolved. The highest authority is considered to be that which is based on reason. Revelations are questioned and never regarded as authentic because they are not acquired by man through reason. Moreover, truths about every aspect of life must be demonstratively provable, conviction depending (according to scientists) not on any certainty intuitively experienced, but solely on facts and arguments demonstrable to others.

A division of the sciences into social and natural sciences indicates an extension of the sphere of sciences beyond the physical world into the field of metaphysics. According to scientific methodology any suggested presence of a Creator in the physical universe is regarded as totally irrelevant. That Allah has willed the universe into existence cannot be demonstrated according to their philosophy whereby Divine Will is ignored and divine sanctions or environments are rejected.

According to Islam, Allah has made man the master of His creation but man must accept those limitations set by Allah. Only then can nature be used for the good of mankind.

In this section an attempt has been made to indicate the extent of these limits and how Islamic values and modern sciences can be integrated.

Syed Ali Ashraf
General Editor

Introduction

Allah has said in the Quran that He has created Man as His Vicegerent and bestowed on him potential mastery over the entire creation. Allah has repeatedly told Man to study the universe, learn the essence of creation and acquire authority. But at the same time Allah has forbidden us to misinterpret and misuse such knowledge. He has warned us that by doing so Man brings upon himself suffering and retribution. The Prophet, peace and blessings of Allah be on him, has instructed us to acquire knowledge. There is no conflict between faith and man's knowledge of the universe and his use of this knowledge for good, provided we do not become, like Satan, arrogant and self-assertive, and start concocting new philosophies to interpret our findings. Islam teaches us humility. It makes us realize that our knowledge is not complete nor will it ever be possible for any man or any creature to have complete knowledge. Only Allah has complete knowledge. It is therefore presumptuous on the part of man to concoct new philosophies about the origin of the universe or the evolution of the species which blatantly contradict Allah's injunctions and assume new hypotheses. It is of course the duty of the scientist to understand and explain creation, the universe and all beings, but whenever his findings and analysis come into conflict with fundamental assumptions stated in the Quran, he should realize that he has not as yet found complete data and that is why he is unable to reach a conclusion that tallies with the statements of Allah.

In the early days of Islam a large number of Muslim scientists carried out much pioneering work but they did not become atheists. Nor did they need to start their work using vague anti-religious hypotheses and to reach anti-Quranic conclusions. Their work did not conflict with religious assumptions because, unlike Christian theologists, Muslim philosophers, theologists and 'ulama did not formulate rigid theories about Nature. Within the framework of Quranic mandates about life and the universe it was possible for these scientists to discover, invent and even theorize. Had not the Muslims become insular as soon as their political decline began their leadership would not have been lost. The 'ulama also decided to save Islam from inroads from various quarters and unanimously gave a verdict against *ijtihad* and all innovation. As a result Muslim thinking gradually declined and the curricula of Muslim universities lost their creative character. A kind of supercilious egoism prevented the 'ulama in Turkey, Egypt, India and other Muslim countries from even trying to understand what was happening in the West. When the Muslims began to lose their authority and to be controlled intellectually by foreign powers, as in Egypt and India, then the newly-learned class woke up. The dominance of modern science misled some Muslim thinkers who attempted to read modern science theories into the Quran. They wanted to prove that all scientific theories were to be found in the Quran. Instead of regarding all scientific theories as tentative they interpreted them as absolute. Fortunately today it is no longer necessary to prove that scientific theories are tentative. Modern scientists themselves say that. It is therefore possible for a Muslim scientist to explore the universe without first of all accepting an anti-Islamic concept as an absolute criterion.

But even if one now considers scientific theories about the universe tentative in nature, one may still not be able to resist the impact of scientific methodology and hence the scientific attitude to life. A scientist relies on reason rather than on feelings or spiritual realizations. It is because of this one-sided approach that specialization in science may generate in students doubts, if not disbelief, in the realizations of Truth by Man's spirit. This may lead to loss of faith. Moreover, it will surely produce an imbalance of personality. The curricula therefore should be so planned that spiritual longings are stimulated and satisfied, emotions are refined and become more perceptive and reason is not allowed to rebel against the supremacy of the spirit.

This goal can be achieved only when (a) scientific knowledge is placed and viewed in the perspective of knowledge about total or whole Man, (b) values as enunciated by Islam are shown to be derived from the concept of the attributes of Allah and are thus regarded as having an absolute norm higher than any scientific knowledge and (c) the value of science and technology is always shown to be properly ascertainable when the relevance and social benefit of science and technology are assessed in the context of human life and man's responsibility to man and God, and his duties to God, society and his own self. The concept of the Whole Man, and the method of assessment, can be taught correctly only when the teaching of Islam is integrated with the teaching of science and technology. This again can be fruitful when humanities and social sciences are Islamized and made an integral part of scientific and technical curricula throughout all stages of education. This is not easy but it is the duty of scholars in all fields of knowledge to formulate a new curriculum and frame appropriate courses for students.

It is with this aim in view that scholars participating in the First World Conference on Muslim Education recommended that scientific fact that may alter with the passage of time should be separated from theory and assumption and that Muslim scientists should abstain from interpreting the Quran through theory and assumption. They also suggested that integrated courses should be planned even if this goal cannot be achieved by science and technology departments working by themselves. It requires Islam-inspired scholars from every branch of knowledge to unite, in a spirit of mutual co-operation, to co-ordinate their efforts and attain this common goal.

Abdullah Omar Nasseef
Editor

Chapter One
Humanistic – Social Sciences Studies in Higher Education: Islamic and International Perspectives

Sayyid Waqar Ahmed Husaini

Sayyid Waqar Ahmed Husaini was born in India in 1937; Ph.D from Stanford University, U.S.A., 1971; Post-doctorate Fellow, Stanford University, 1974-75, Associate Professor, College of Engineering, King Abdul Aziz University, Jeddah, Saudi Arabia; Senior Lecturer, M.A.R.A. Institute of Technology, Malaysia, 1972-74; United Nations expert Consultant for U.N.E.P., U.N.E.S.C.O., W.M.O.; General Secretary, Association of Muslim Scientists and Engineers, U.S.A. and Canada, 1974-75; General Secretary, Asian Environmental Society, 1972-74. Publications include: *Principles of Environmental Engineering Systems Planning in Islamic Culture: Law, Politics, Economics, Education and the Sociology of Science and Culture* (1971, Stanford University, U.S.A.).

Introduction

This paper reviews the impediments in, and the prospects and plans for, integrating Islamic ideology with humanities, social and natural sciences, and technical subjects in Muslim higher (university) technical education. Contemporary Western and Marxist systems of higher technical education are reviewed which have successfully integrated the two ideologies or two world-views in the teaching of humanistic – social sciences and technical subjects. The characteristics of the early medieval Muslim education system which integrated Islam with science and learning, and those characteristics which integrated the late medieval system with modern Muslim higher education and created a schism between rational and Islamic studies, are discussed. A realistic plan of action to integrate Islam in contemporary Muslim technical education is presented in the context of the continuing efforts at the College of Engineering, King Abdulaziz University.

University engineering education is used as an example of higher technical education. The arguments, conclusions, and plan of action presented are considered to be basically valid and applicable to education in all other, higher or vocational, scientific and technological studies in the geophysical sciences, agriculture, medicine and their allied sciences. The term "technical" is used in the broadest possible concept for all natural and applied sciences, and technology.

The terms *ulum aqaliyya* (rational sciences) and *ulum Shariyya* (Islamic ideological sciences) are used but differ in some ways from their classical meanings: The *ulum Shariyya* are disciplines, or parts of disciplines, which are predominantly based on or derived from Islamic ideology, ethics, and value-judgements. Thus jurisprudence, welfare economics, science policy, and philosophies of natural or social sciences can be considered *ulum Shariyya*. The *ulum aqaliyya* are disciplines, or parts of disciplines, which are derived from pure reason, experience, and experimentation. The sources, methods, and subject matter of these sciences, or their parts, are not directly within the scope of the Quranic revelation and Islamic ethics. Linguistics (including Arabic), positive economics or positive law, geophysical sciences, engineering, medical sciences, and technology are examples of *ulum aqaliyya*. Generally speaking the humanistic and social sciences belong to the category of *ulum Shariyya*, and the geophysical and biological sciences and technology to that of *ulum aqaliyya*. Succinctly, the relations between the *Shariyya* and *aqaliyya* disciplines, or components of a discipline, will be as complex and intertwined as the relations between reason and the Revelation, facts and values, faith and experience. Indeed, the formulation and solution of these epistemological, philosophical, and axiological controversies is, perhaps, the greatest single achievement of the medieval Muslims, and an accomplishment of the intellectual mission of mankind.[1]

The Structure of Higher Technical Education and Training

Higher education, technical or otherwise, should have two basic components:

1. Pure and applied sciences and mathematics; linguistics; and

technology or technical specialization (including arts, crafts and vocational skills). In the Islamic philosophy of knowledge and education, these fall into the categories of *ulum aqaliyya*, and the "socially obligatory" (*fard kifaya*) knowledge. Communication skills in the language of a leading industrial country should be included in this division (e.g. English, Russian, French).

2. Humanities and social sciences, whether or not integrated with natural sciences and technology as outlined above. These are included in the categories *ulum Shariyya*, and the "individually obligatory" (*fard ayn*) knowledge. Knowledge of Arabic, though it is a "rational science", must be considered obligatory for all Muslims.[2]

Higher education must have these two components throughout the later stages of specialization. However, the emphasis may shift to one or the other component depending on whether the major specialization is in *ulum aqaliyya* or *Shariyya*. Such a system of education in which the rational and the ideological components are developed and organized into a complete whole, is an *integrated system* except for one additional requirement. Acquisition and utilization of knowledge must accompany a development of those virtues and wisdom which lead to success and happiness in the hereafter. "That is the supreme achievement".[3]

The classical and early medieval Muslim educational systems were, indeed, Islamic integrated technical education systems. They produced the Islamic, integrated, "scientist-philosophers". These learnt the Quran by heart and they knew the *shariah* and the exegesis of the Quran. Most of them were non-Arabs who mastered Arabic, who learnt the necessary humanistic-social sciences and practised virtue and exuded wisdom, and yet, or because of all this, they were the greatest scholars in natural sciences and technology of their times.[4] From the end of the fifth/eleventh century, the Islamic integrated technical education system began to disintegrate. First the integrity of the educational system was destroyed when science, the humanities (philosophy), and social sciences were excluded from the curricula. Higher Education curricula were dominated by "Islamic" (i.e., sectarian, theological) studies, spoken Arabic and grammar, and literary studies.[5] In the pre-modern period until today, while the curricula of the Shariah colleges have continued as such with some modifications, a dual system has been developed with a parallel system of technical education, humanities, and social sciences, each system excluding the other component. The break-up of the integrated educational system is the primary cause of the decline of both Muslim science and Islamic ideology. A schism was created between the Muslim "traditionalists" and the Muslim "modernists". Consequently politico-cultural heretic movements also developed, often in collaboration with the enemies of Islam against fellow-Muslims, causing the decline of Muslim political power and influence, and contributing to the decline of Islam both as a cultural force and as a civilization.

The pre-requisites for a contemporary rejuvenation of Islamic civilization are the restoration of Islamic humanistic-social sciences studies in technical education, and the restoration of the study of natural sciences and technology in humanistic-social specializations, particularly in the faculties of Shariah and Islamic Studies. The Faculty of Humanities and Arts, and the Faculty of Shariah and Islamic Studies could with advantage become one and the same.

The restoration of the integrity of the education system through the introduction of the two components, each in complementary proportions, is an Islamic epistemological necessity. One cannot be ignorant or independent of the *ayat Allah* in either natural sciences and technology or in humanities and social sciences. This is the essence of the "Two-Book" concept in the Quran of the revelation of God's will.[6]

The Objectives, Goals, and Structure of Humanistic-Social Sciences Studies

The Objectives

Humanistic social-sciences studies, and their ideological orientation towards Islam, must have the following major objectives:

(1) The development in thought and deed of the personal and social ethics of Islam for a successful life in this world and the hereafter;

(2) an appreciation of the social, economic, political, cultural, philosophical, national and developmental factors in the analysis of scientific-technological problems and solutions;

(3) an understanding of our national and regional social environ-

ment, and our micro-biosphere, which influence, and are influenced by, science and technology;

(4) a development of personal and group attitudes and abilities that could enable a student to give his technical education and training, and research and development efforts, a situational orientation; that is, a focus on application, development, field and practical work relevant to national and regional developmental problems;

(The above two objectives require originality and resourcefulness, i.e., an innovative mentality (*ijtihadiyya*) and creatively imitative (*taqlidiyya*) mentality, which are distinct from pure or blind imitation (*taqlid mahd*). This will enable our scientists and technologists to apply technology correctly. The consequences of misapplication of modern technology are all too obvious in developing countries. Some Muslims will be able to learn from the self-criticism of industrialized societies inspired by the growing recognition of a world environmental crisis.)

(5) a self-realization of the historical and contemporary behavioural cultural patterns of Muslims which caused the decline, and the retardation, of our scientific and technological development. Also the self-realization of those ideal cultural patterns of Islam, historically manifested in the Muslim civilization of the first/seventh to fifth/eleventh centuries, which developed a leading and an excellent system of science and civilization. This objective is also intended to inculcate the Islamic rational or scientific method so that science and technology become readily acceptable to ourselves and our cultural ethos. They should not remain, as they often do, an appendage and an artificial civilization alien to men's minds and their social behaviour.

The basic goals of these objectives are:

(a) to integrate science and technology with the ideology and cultural ethos of Islam, and to place them in correct perspective to the history and sociology of Muslim science and technology; and

(b) to integrate Islamic ethics and ideology with the humanities and social sciences, and to give these too a sense of continuity in the history of Islamic thought.

Curricula Structure and Functions

Four major elements or functions of a balanced curriculum structure for humanistic-social sciences studies in technical education are discussed. These are already recognized and applied in industrially advanced countries, and even in those less developed countries but which are advanced in ideological self-consciousness.

(1) Indoctrination and Acculturation

Humanistic-social sciences studies in particular, and science and technology in general, inculcate different ideologies or world-views such as the Islamic, the secular and agnostic, the Western, and the marxist-Leninist. Man and his technology do not exist in a values vacuum. Humanistic-social studies are the primary vehicle in imparting value-orientations and behaviour-orientations for the acculturation and socialization of adults. These processes occur within an ideology or a socio-cultural system which may be religious, non-religious, or anti-religious. All courses, technical or non technical, must provide indoctrination and acculturation, though courses offered to perform certain functions may not have this as a primary focus.

(2) The Humanistic Function

Technical specializations, through fragmentation of human knowledge and experience, promote a converging mental attitude. The humanities and social sciences provide the corrective to this tendency by exposing the student to those disciplines which enable him to perform his personal, familial, social and civic functions with well-informed understanding. This humanistic function implies that such humanistic-social sciences courses do not have direct relevance to technical specialization or for professional efficiency.

(3) Social Responsibility

Scientists and technologists are involved in multivariate systems which affect morality, and the life, health, and welfare of man. They should be educated to assume leadership in policy-making, decision making, and management of small enterprises as well as involvement in government. The world wars of this century as well as the environmental crisis have reinforced the need to break into the isolation of the scientist and technologist within the narrow confines of his specialization, and to hold him responsible for the wider consequences of the application of science and technology. They are the main agents of change in socio-economic development and industrialization in the less developed countries. These functions of the scientist and technologist require increasingly significant new roles for humanistic-social studies, particularly the "integrated" or multi-disciplinary studies at the interface of science and technology with humanities and social sciences.

(4) The Instrumental Function

The relations between man and machines, the social environment, and the processes in the production of scientific and technological goods have given rise to new applications of humanities and social sciences. To improve the professional efficiency of the scientist and technologist, new techno-humanistic and techno-social sciences disciplines have been created. Industrial psychology, industrial sociology, industrial law, industrial economics, and industrial management are just a few of the inter-disciplinary subjects which are favoured for their direct, positive value for the practising scientist and technologist. They have an "instrumental" or "practical" value. The development of Islamic economics, for example, is a big intellectual challenge for Muslim economists. The development of further specialized disciplines which are multidisciplinary, such as industrial economics or engineering economics, is a higher challenge.

The humanistic-social sciences curricula in technical education must have a well-planned series or sequence of courses in each of the above four functional areas.

The Predicament of Humanistic-Social Sciences Studies in Muslim Technical Education Systems

Technical education may be imparted through a world-view (1) formed by an ideology or religion, or (2) in a purely rational and secular context which is supposed, erroneously, to be independent of any ideology. Western and Marxist countries have succeeded in integrating Western and Marxist ideologies with the methods and contents of their technical education and training systems.

The Muslim world consists of independent Muslim-majority countries, as well as subjugated Muslim-majority regions or countries, and the minority Muslim communities living in non-Muslim majority countries.

The single most important characteristic common to all Muslim technical education systems, moderated, more or less by concessions to Islamic cultural influences and sensitivities, is their domination by either Western or Marxist world-views. One such concession is to misrepresent education systems to ourselves as being purely rational and independent of any ideology. We do this also, perhaps, for pragmatic reasons. The dominant model of these systems comprises three parts of the curricula structure:

(1) Pure and applied sciences and mathematics, and technology or technical specialities.

Technical disiplines are taught in "Muslim countries" in a purely rational context devoid of all values and ideological interpretations of cosmology. The absence of any reference to non-Islamic ideologies or other religions is secularism. This is an antireligious and negative ideology and cannot be considered a positive Islamic frame of reference. Furthermore, since the books, the non-Muslim instructors, and the Muslim instructors who are themselves students and practitioners of the Western or Marxist educational systems, are all products of at least non-Islamic origins, even these rational sciences and technology embody the Western or Marxist cosmological doctrines.

(2) Techno-humanistic and techno-social sciences, or inter-disciplinary or "integrated" studies.

These "applied" humanistic-social sciences mostly fulfil the instrumental function of humanistic-social sciences studies. In the teaching of these subjects the available Western or the Marxist literature, deceptively called "modern" and "socialistic" respectively, is used. Their plagiarized versions in Western or local languages, if they do exist, might be qualitatively inferior but in ideas and content they might be mostly the product of uncritical and blind imitation. To this category belong the works of industrial management, engineering economics, industrial psychology, economics of public works planning etc.

(3) Humanities and social sciences

These disciplines or subjects belong to the areas of the curricula structure described above as the "humanistic function", "social responsibility", and "indoctrination and acculturation".

Even in avowedly Muslim countries, social sciences are taught in a Western or Marxist ideological perspective, and the humanities are taught from a Muslim perspective. The Islamic humanistic sciences may or may not be a required or compulsory part of technical education curricula. Even in universities in which Islamic humanistic-social science studies are a part of technical education curricula, there may rarely be any relevance in these courses to problems at their interface with science, technology, industrialization, and socio-economic development. They are taught by graduates of the traditional colleges of Shariah or modernist departments of Islamic studies in Muslim or Western countries. These are mostly traditional humanistic courses in Muslim history, theology, beliefs and practices, exegesis and dialectics, apologetics and polemics, Muslim historical romanticism, etc.

The technical education curricula of secularized Muslim countries often include neither Islamic nor non-Islamic humanistic-social sciences studies. This is partly due to the misconceptions among Muslim technical educationists and professors concerning the positive role and necessity for humanistic-social sciences in technical education and technological development. Humanities and social sciences are considered a waste of time, of low priority, or irrelevant for education in science and technology. These misconceptions are common among secular and modernist Muslims as well as among the "good Muslims" who conscientiously practise the "five pillars" of Islam. A second reason is that these Muslim technical educationists and professors are products of a colonialist education system whose aim was to produce narrow technical specialists incapable of social understanding, involvement, and development. Third, these "educationists" and professors are most often utterly ignorant of the history, philosophy, and methodology of Islamic technical education as well as that of the contemporary industrialized countries where they themselves obtained higher education in narrow technical specializations. One could find among them, for example, members of professional engineering associations of the U.S.A., and contributors to their journals. Rare, indeed, are found any Muslim members in the American Society for Engineering Education, or Muslims aware of UNESCO activities and publications concerning humanistic-social sciences studies in technical education. The fourth reason, and perhaps the most significant for the lack of awareness among Muslims of humanistic-social sciences in technical education, is the dearth or non-existence of any relevant and meaningful Islamic literature in the humanities and social sciences. We also lack instructors who are qualified or at least willing to take up this intellectual challenge.

The third reason mentioned above is reviewed below before passing to the fourth and the presentation of a plan of action and a case study.

The Teaching of Humanistic-Social Sciences Studies in Higher Technical Education in Western and Communist Countries

This section gives a brief survey of the status and trends in the teaching of humanistic-social sciences in higher technical specialist engineering education in the U.S.A. and the U.S.S.R. representing the Western and Communist countries. Case studies of other countries can be found in miscellaneous literature.[8] No studies have been found in English literature concerning any Muslim country.

The U.S.A.[9]
The American Society of Engineering Education (ASEE), formerly the Society for the Promotion of Engineering Education (SPEE), has since 1918 been sponsoring special studies in humanistic-social sciences. The ASEE has consistently recommended that about 20% of the undergraduate engineering curriculum should be devoted to the humanities and social sciences. The Engineers' Council for Professional Development (ECPD), which accredits engineering programmes of colleges in the U.S.A., requires about one-sixth to one-eighth of the degree programme to be in the humanities and social sciences. But the subjects fulfilling the "instrumental" function are usually not included in this quota. Data from nation-wide surveys have shown that the humanistic-social sciences component of the curricula (excluding linguistics) was in the following proportions:

29·1% in 1870, 16·7% in 1923, 16% in 1939, 14·5% in 1949, 14·7% in 1955, 17% in 1959, 18·3% in 1961, 15% in 1968, and 17% in 1973.

A 1963 national survey showed that engineering college curricula included social sciences studies in the following descending order of preference: economics, history, political science, sociology, psychology, geography, anthropology, law, and public administration. The Humanities subjects included: English, speech, literature, languages, philosophy, religion, and the fine arts.

A 1940 SPEE Report recommended that engineering curricula should consist of two "stems"; the scientific-technological and the humanistic-social sciences. Courses in the two stems were not interrelated or "integrated", and any connections between the stems were unplanned and accidental. However, a 1968 ASEE Report proposed that the two separate "stems" should be "integrated" for greater communication between the technical and the non-technical disciplines. However, a 1973 ASEE Report disclosed that a large number of colleges continued to have a humanistic-social science stem as a separate, non-technical, part of the engineering curriculum. These courses were taught in different departments of the faculties of humanities and arts. However, integrated courses, taught jointly by engineering and humanities of social sciences faculties, increased from 14% in 1968 to 76% in 1973.

Humanities and social sciences are taught in the U.S.A. in the national and institutional framework of American culture and technology, the Western ideologies of political democracy and capitalist economy, and in general within the ethnocentric perspectives of Judeo-Christian Western civilization. The teaching of history, for example, might consist of a year-long series of courses in the History of Western civilization. The content of the courses and books used present a typical Western view of the history of science and technology, history of ideas and philosophical movements, the glorification of Western heroes and intellectuals. This history jumps over the millennial Dark Ages between the ancient Greeks and Romans, and the Renaissance and the Reformation. The attitude towards the medieval Islamic origins of Western science and technology, philosophy and theology, education and culture, before and during the Renaissance and the Reformation, is one of a conspiracy of silence and systematic falsification of history. The product of such a Westernized system of humanistic-social sciences studies is the typical self-assured, ethnocentric, and arrogant Western man. He is indoctrinated to believe in the originality, uniqueness, superiority, universality, and the inevitable triumph of Western-Christian civilization.[10]

The U.S.S.R.[11]

The purpose and philosophy of the Soviet education system is to instil loyalty towards Communism, belief in its superiority over other systems, and to inculcate the Marxist-Leninist world-view and patterns of behaviour.

Nearly half of all university level students study engineering. About 13% of the scheduled time in all technical institutes is assigned to humanistic-social studies excluding linguistics. A brief description of the subjects taught at Soviet engineering and other higher technical institutes is given below. The numbers denote class-contact hours during a year.

First Year

1. History of the Communist Party of the Soviet Union (100 hours)
Treats the subject as a most important part of the universal history of mankind. Deals with past and modern history, sociology, philosophy, political economy, and ideological aspects of society. Relations between the ideological, political, and scientific-technical activities of technical specialists are emphasized.
2. Marxist Ethics (24–32 hours)
Provides moral education through discussion on freedom and the necessity in ethics, moral ideals, professional ethics, moral relations in Soviet society.

Second Year

3. Marxist-Leninist Philosophy (70 hours)
Demonstrates that dialectical and historical materialism provides the methodological base for the teaching of all other social and scientific subjects.
4. Marxist-Leninist Aesthetics (24–32 hours)
History of aesthetic thought. Basic propositions of Marxist aesthetics and critique of contemporary aesthetic theories.

Third Year

5. Fundamentals of Scientific Atheism (24–32 hours)
Inculcates the atheistic world-view of Marxism. Origin of religion, attitude of the Soviet State towards religions (such as Islam and Christianity), religious organizations, and ways of combating them.

Students given atheistic education also in subjects such as physics, mathematics, chemistry during lectures and practical studies.

6. Political Economy (110 hours)

Departments of Political Economy in higher technical schools prepare this course which is based on programmes of the Soviet Communist Party prepared for the successful building of Communism. A profound knowledge of socialist economics, political economy and scientific planning, is provided.

Fourth Year

7. Fundamentals of Scientific Communism (70 hours)

The purpose of this course is to show the prospects for development of mankind on a new and higher economic, technical and social level through Communism. Deals with origins and development of scientific Communism, and its creation by Marx, Engels and Lenin. Critique of liberal democracy, national liberation movements and non-Marxist socialism. Strategy and tactics of Communist Parties.

Discussion

One finds considerable variation in the proportion and content of the humanistic social-sciences component in the technical curricula of non-Communist industrialized countries. This is due to the variations in the national systems of higher and pre-university education and, most significantly, the role expectations of engineers and other technically qualified persons in the national economy and industry. However, the ideological content and socio-cultural orientation is unmistakably according to their own perceptions of civilization and the national self-image.

The humanistic-social studies component of technical education in the East European Communist countries is, scholastically and in structure, almost identical with that of the Soviet Union.[12] In the Peoples' Republic of China, students spend 75% of their time in class in scientific-technological studies, and 15% in Marxist-Leninist humanistic-social studies influenced by the thought of Chairman Mao. The study of military affairs, and work in factories and on the collective agricultural farms, take up 10% of curricula time.[13]

UNESCO studies have shown that the teaching of the humanities and social sciences in the technical education of the non-Communist developing countries is very much neglected and downgraded. This situation however seems to be improving now. The reason most often given in the past has been that in the present stage of scientific-technological development the country cannot afford to dilute its technical curricula through the introduction of humanistic-social sciences studies. The time allocated to the humanistic-social sciences in U.S. engineering education a century ago, and throughout the thirteenth/nineteenth century, perhaps since their first engineering college was established in 1802, was higher than it has been in recent decades. That was the period of the birth and rapid growth of its technological and economic development. Its emphasis on such studies have never diminished though engineers are working in an increasing number of technical specialities as well as in managerial positions in industry and government. Technical education in the rapidly developed Communist countries shows too their untiring efforts to mould and remould man through Marxist-Leninist ideological studies for industrial and socio-economic development. Indeed, there is a vast literature on the sociology of culture and development that proves that technological and economic developments were caused by favourable religious and ideological systems. The early post-Quranic history of Islamic science and civilization, and a central theme in the incontrovertible Quran, also prove this point. Sorokin, the sociologist of science and civilization, made a statistical study of the indicators of growth in science-technology (the material or technological culture) and humanities-social sciences (the non-material or ideological culture). He found that in Islamic (A.D. 700–1300) and Greco-Roman and Western civilizations (B.C. 600–A.D. 1900), and in individual countries, the indicators for creativity and growth in the technological and ideological cultures grew or declined simultaneously.[14]

Technical education in Muslim and other Afro-Asian countries has been developed in recent decades on the model of, and with assistance from, similar institutions in the U.S.A. and the U.S.S.R. Nevertheless, the traditions and trends of these in humanistic-social sciences studies in technical education have had no demonstrable effect on the design of the curricula in the countries concerned. The negative influences can be seen in the content of the curricula, the humanistic-social sciences as well as the scientific-technological components. The intellectual and political elites of the Muslim world have been too weak in will to assert their economic, politico-cultural and ideological independence. The *ijtihadiyya* mentality, and the institutional infrastructures that facilitate its development, have been conspicuous by their absence or have been unproductive. Other reasons have conspired to make Muslims passive, ignorant, and even eager imitators of the Western and the Marxist humanities and social sciences. Western ideology is considered to be "modern" and "scientific". Marxist thought is considered to be "socialistic" and "progressive". Both are presented as universal ideological cultures. Either may be taken up by Muslim intellectuals and ruling elites, who are willing to accept ideological compromises and cultural bondage as the price of wealth, power, and prestige. They have to fight against "traditionalist" or "reactionary" fellow-Muslims under the banner of modernism or socialism. The identification of Islam with modernism or socialism makes the Islamic ideological struggle unnecessary, and prepares the ground for the acceptance as indigenous of what is extraneous to Islamic ideology and cultural ethos. Thus "modernism" has become a camouflage for secularization and Westernization. "Socialism" has become a disguise for Marxist-Leninism. They have been effective intellectual ploys for cultural self-capitulation and successful propagandist slogans of cultural imperialism.

Whither do we go and what shall be our endeavour?

A Plan of Action

The medieval Muslims made epistemological breakthroughs, and discovered the fundamental principles of Islamic educational philosophy. These must be recapitulated and implemented in a dynamic *ijtihadiyya* spirit. Some of the basic issues and principles of Islamic philosophy and structure include: the classification of knowledge by sources and by methods of knowing, and clarification of the domain, subject matter, goals and inter-relations, for example, of the *ulum Shariyya* and *aqaliyya*; individual and social criteria for the pursuit of education, research and development; recognition of the sources of knowledge in curricula design and content; inter-dependence and hierarchy of virtue, values (ideology) and facts (reasons), etc. Thus Islamic humanistic-social sciences must be an integral and thoroughly integrated part of technical education.

The first and most important task is to begin the integration of Islamic ideology with the humanities and social sciences. This is basically the responsibility of the specialists in the following disciplines: law, political science, economics and sociology. This cannot and should not be done by Muslim activists, public leaders and preachers. But who are these "specialists"? We are aware of the ignorance or the insufficient knowledge of Islamic ideology among professors of our secularized faculties of the humanities and arts. Similarly, our professors of the Shariah faculties and the traditional "scholars of Islam" (*ulum al-din*) are either ignorant or have insufficient knowledge of the (secularized, Westernized, or Marxianized) social sciences and humanities. If they have humility, faith, will and ability, both these groups of "half-scholars" can perform the great responsibility of developing the Islamic humanities and social sciences. One institutional arrangement that should be attempted is to abolish the *separated* faculty of humanities and arts, and the faculty (or department) of Shariah and Islamic studies. These should be merged into one faculty with various departments on the basis of the specialization or interest of a particular faculty member in a particular discipline. The "re-training" of our "half-scholars" in the areas of their relative ignorance or insufficient knowledge should be undertaken by various methods such as self-education and research during a period of sabbatical leave, formal education through a Master's degree or a second doctorate, rubbing shoulders in humility with colleagues in the "merged" departments, etc. The Islamic scholar-activists should be invited to universities as visiting professors as well as part-time or full-time students.

The body of knowledge developed in this way will serve the needs of technical education by providing textbooks and monographs for courses serving the "humanistic function", and "indoctrination and acculturation" in particular. They will also provide the foundation for courses serving the other functions of humanistic-social sciences curricula in technical education.

A second task is the integration of Islamic humanistic-social sciences with science and technology. The faculties and departments of science-technology specializations (for example, physics, biology, agriculture, medicine, engineering) should develop a body of knowledge specializing in Islamic history, philosophy, and sociology of these science-technology specializations as well as of science and technology in general. A course in the Islamic (and comparative) history and philosophy of mathematics should be a requirement for students specializing in mathematics, for example. Such courses would serve the "social responsibility" function of technical education.

A third task is the integration of Islamic ideology with techno-humanistic and techno-social sciences disciplines, or multi-disciplines. This can be done by teams, collaborating in teaching as well as in research and publications, comprising faculty and non-academic people who are specialists in technical subjects and the Islamic humanities and social sciences. This body of knowledge and expertise will serve the "instrumental function" of technical education.

The responsibility for implementing this general plan of action falls on many shoulders. The institutional structure can be of various types and these institutions can be created and supported by various means. Individuals in the community or in a university faculty in their spare time, faculty members released from their teaching and administrative duties for part-time or full-time research, and students pursuing higher studies, in the Muslim-majority as well as the Muslim-minority countries, should all perform this social obligation (fard kifaya). Islamic political, activist, philanthropic, and propagation organizations must realize that their efforts will most likely be doomed to failure unless we have literature on Islamic humanities and social sciences comparable to similar Western and Marxist books, journals, etc.

A special responsibility rests on the universities and governments of Muslim-majority countries, and their Islam-conscious elites. Islamic research institutes, Islamic conferences, ministries of religious affairs, and organizations concerned with the construction of large mosques and monuments, should be vehicles for the development of Islamic humanistic-social science disciplines. While it is proper to aim at a "big push" to achieve this goal, the Islamic strategies of gradualism (*tadrij*) and easy, small compromising steps (*taysir*) might be preferable in most circumstances. Each department of a faculty or college in a university might create one or more positions for "research professors" in an Islamic humanistic-social science specialty. Research and publication are better fostered when there are opportunities to teach and to use their output.

The Muslim-minority Western countries do provide, in some ways better than the Muslim-majority countries, opportunities for research, publication, and even the teaching of Islam in humanities and social sciences. One of the biggest impediments in utilizing these opportunities is the lack of the will of the Muslim student or professor in a Western university who may not have an Islamic ideological orientation. These are often involved in building up personal careers quite unnecessarily through the conscious assimilation of Western intellectual and applied research ideas. The regaining and re-direction of these lost souls among Muslim immigrants in the West could provide a tremendous boost to the cause of the Islamic humanistic-social sciences.

The Muslim-minority communities in non-Muslim countries must realize that the spread of Islam in their countries requires provision through the Islamic humanistic-social sciences of the intellectual and ideological foundations for universal social justice. This in turn requires the willing acceptance by non-Muslims of the Islamic humanistic-social sciences through their proven superiority. Meanwhile the Muslim-minority communities must try to ensure that Islamic humanistic-social science courses are included in the approved list of electives acceptable in the technical education curricula of universities in their countries.

NOTES

1. For detailed discussions on the adaptation of Islamic educational philosophy to engineering education, and an international survey of humanistic-social studies in engineering education, see, Husaini, S. Waqar Ahmed, *Principles of Environmental Engineering Systems Planning in Islamic Culture; Law, Politics, Economics, Education, and Sociology of Science and Culture* (Programme in Engineering-Economic Planning, Report EEP-47; Stanford: Stanford University, California, 1971); and Husaini, S. W. A., *Humanistic-Social Studies in Engineering Education; International and Islamic* (To be published)

2. For medieval Islamic epistemology and education system, see al-Ghazali, *The Book of Knowledge*, being a trans. with notes of the *Kitab al Ilm* of al-Ghazali's *Ulum al-Din* (Lahore: S. M. Ashraf, 1966); Muhsin Mahdi, *Ibn Khaldun's Philosophy of History* (London: G. Allen & Unwin Ltd, 1957); Ibn Khaldun, *The Muqaddimah: An Introduction to History* (Trans. F. Rosenthal, Bollingen Series XLIII: 3 Vols: 2nd ed. Princeton: Princeton Univ. Press, 1967); Munir-ud-Din Ahmed, *Muslim Education and the Scholars' Social Status* (Zurich: Verlag Der Islam 1968).

3. Quran 9:72; 5:119; 45:30; 6:15f; 9:111; 10:62–64.

4. George Sarton, *An Introduction to the History of Science* (3 Vols. in 5: Baltimore: Williams and Wilkins Co., 1927–1948); Sayyed H. Nasr, *An Introduction to Islam Cosmological Doctrines* (Cambridge: Harvard Univ. Press, 1964); S. H. Nasr, *Science and Civilization in Islam* (Cambridge: Harvard Univ. Press, 1968).

5. Mehdi Nakosteen, *History of Islamic Origins of Western Education A.D. 800–1300* (Boulder: Univ. Colorado Press, 1964) Page 42.

6. S. A. Latif, *Bases of Islamic Culture* (Hyderabad, India: Institute of Indo-Middle East Cultural Studies, 1960); Ismail R. Al-Faruqi, "Science and Traditional Values in Islamic Society", in *Science and Human Condition in India and Pakistan*, ed. W. Morehouse (New York: Rockefeller Univ. Press, 1968).

7. For example, UNESCO, *Social Sciences and Humanities in Engineering Education* (Studies in Engineering Education 2; Paris: UNESCO Press, 1974), and UNESCO, *The Teaching of the Social Sciences in Higher Technical Education: An International Survey*, ed. J. Gould and J. H. Smith (Paris, UNESCO, 1968).

8. *Ibid*, and D. W. Sallet, "Education of the 'Diplôme Ingénieur' " *Engineering Education*, V.59 (June 1968), 1105–1106: Sir O. A. Saunders, "Trends in Engineering Education in Western Europe", *Engineering Education*, V.61 (Dec. 1970) 279–82; A. C. Gross, "Selected Aspects of Engineering Education in Canada", *Engineering Education*, V.59 (June 1969), 1110–1112.

9. The UNESCO publications in f.n. 7 and ASEE, *Final Report Goals of Engineering Education* (Washington, D.C.: ASEE, 1968): ASEE, "Liberal Learning for the Engineer", *Engineering Education*, V.59 (Dec. 1968), 303–42: ASEE, "Liberal Learning for the Engineer: An Evaluation Five Years Later", *Engineering Education*, V.65 (Jan. 1975), 301–24.

10. Max Weber, *The Protestant Ethic and the Spirit of Capitalism*, trans. T. Parsons (New York: Scribner's Sons, 1952).

11. The UNESCO publ. of f.n. 7, and N. P. Kuzin, *Education in the U.S.S.R.* (Moscow: Progress Publishers, 1972): A. G. Korol, *Soviet Education for Science and Technology* (New York: Technology Press and J. Wiley and Sons, 1957); H. W. Butler, et al, "Engineering Education in the Soviet Union", *Engineering Education*, V.63 (Jan. 1973), 276–80, 289.

12. See above f.n.7.

13. T. Durdin, "Teacher 'a New Man' after remoulding by Cultural Revolution", *The New York Times*, Apr. 30, 1971; "Engineering Education in China", *Engineering Education News*, V.1, No. 8, Feb. 1975, P.1.

14. P. A. Sorokin, *Social and Cultural Dynamics* (4 Vols. New York: American Book Co. 1937–41), IV, 323–88, 145–96.

Chapter Two
Scientific Education in Muslim Countries — Principles and Guidelines

Atur-ur-Rahman

Atur-ur Rahman was born in 1942. He is Co-Director, Post-graduate Institute of Chemistry, University of Karachi, Pakistan. He is also a member of A.R.I.C., Chemical Society of London and life member of the Cambridge Chemical Society. His publications include a book, *Biosynthesis of Indole Alkaloids*, and more than 35 research papers in leading international chemical journals.

The problems associated with scientific and technological training are basically two-fold. The first and foremost is the almost total dependence of the Muslim world on Western science and technology. We must shrug this off and develop an independent dynamic infra-structure in the shortest possible time. The second closely related problem is how to absorb these technological advances into the Islamic rationale without being affected by the materialistic philosophies that emanate from the West. In order to counter the ulterior effects of the mad rush towards technological advancement, especially in the oil-rich Arab states where money is now being poured out to buy Western science and technology, it is necessary that students of science in the Muslim world should have a clear understanding of the various Western materialist philosophies and their pitfalls.

Materialism is an outlook which hinges on the belief that matter is the be-all and end-all of the universe. It lays down that there is no more to nature than can be analysed in purely physical terms, and a search for an understanding of the spiritual world is a vain one as materialism does not believe in its existence. Since the study of the sciences has been concerned with the exploration of the properties of matter in its various forms, there has been a close connection between science and materialism. It is important to understand the falsity of this relationship before we go any further. Since the sciences are concerned with the study of matter, the results of scientific experiment provide a deeper insight into the structure and properties of various material substances. This understanding is purely relative to the state of knowledge of the subject at any given time and it may undergo drastic changes, even complete reversals, with the accumulation of further experimental data. An example is the nature of the atom which was thought to be completely understood only half a century ago, but with the discovery of numerous "fundamental" particles of which each atom is composed the whole of Physics is in a turmoil. So "truths" of yesterday become the "semi-truths" of today and "falsehoods" of tomorrow. But even if the "whole truth" about any physical substance or conglomeration of substances were to become known (an impossibility of course) it would provide evidence no more and no less than that relating purely to the nature of the material substance being studied. To derive conclusions from such data about the entirely different spiritual world would be like trying to derive information about the structure of nucleic acids from the braying of a donkey. There can be no possible logical basis for projecting data obtained from the study of matter to areas completely unrelated to matter.

It is unfortunate that a hazy perception of the precise limitations within which scientific studies operate has led to the growth of materialistic philosophy. It is most important that throughout the Muslim world this haziness should be dispelled so that the study of sciences can be vigorously pursued without fostering materialistic philosophies. The need for this becomes all the more acute in nations such as Saudi Arabia where attempts are being made to pack centuries of growth into decades. The danger that in this mad rush towards so-called "progress", moral values may fall by the wayside and give way to a new class of "educated" but non-religious "Muslim" people, is very real and needs to be tackled without delay. This is of course not to underrate the importance of achieving parity with, and ultimately superiority to, the West in sciences and technology. This must, however, be done after taking all necessary precautions to prevent the poison of materialistic dogma from seeping into our educational institutions. The technique must remain a technique and must not be allowed to be taken as a philosophy of life.

It is important also not to be dazzled by Western scientific achievements which may appear great in some eyes but when examined more minutely lose their initial lustre. Let us take space travel, for instance. Man's journey to the moon and the recent explorations of Mars appear to be fantastic achievements. When considered, however, in the wider perspective of the overall size of our universe the distances travelled are indeed trivial. In fact it becomes clear that Man in the entire life span of his species on earth will never be able to explore but a minute portion of the entire galaxy. Similarly when one looks at the other extreme of the universal scale, the world within the atom, one also finds it full of unexplored wonders. In fact it becomes clear with even a little depth of thought that man will never know "everything about anything". To me as an organic chemist, the beauteous wonders of the world within the living cell never lose their fascination. Thousands of chemical reactions take place every second, each designed for a specific purpose, the vast majority of which we do not understand at all, proteins being synthesized from various amino acids to provide hairs in one array of combination, nails in another and enzymes in a third. The chemistry of memory and that of genetics are other equally fascinating areas. And all these wondrous processes occur without any deliberate effort on our part.

Our probing intelligence, at whichever level of magnification, can only be overawed by the wonderful organizational beauty of the natural processes. It would be beyond the scope of this article to discuss the forces at work behind the organizational paraphernalia of nature. I can only refer the reader to Surah "Al Rahman" of the Quran where detailed reference has been made to some of the miracles around us.

It is often assumed by scientists themselves that the study of science is purely a pursuit of truth, divorced from the moral, philosophical, political, economic and religious implications that guide other human beings. This subject is thought to be the only determining factor in planning new experiments, and the scientist happily carries on in this "sealed and insulated" environment oblivious and often unmindful of the impact that this "search for truth" will have on his fellow human beings. Clearly the canvas of a scientific worker is as wide as his mind and he has a large number of areas of research to choose from within his own specialized field of interest. This choice is determined by the interest of the investigator, the facilities available, and in more recent years by the funds available through various national and international funding agencies to support his research. This last factor is in fact the "control box" through which any country can, after determining the priority areas of research, manipulate the extent of effort to be put into each area. The indiscriminate use of the funding "control box" can have a serious, constricting effect on research. However, its use is clearly necessary, especially in the "catching up stage" of our scientific development where intensive efforts have to be made in certain specific areas of research. A study of the pigments in the wings of butterflies, for instance, may be of considerable academic interest in the knowledge of their structure but it will hardly contribute to the achievement of technological independence. One can of course put the argument, and indeed a very powerful one, that such an application of funding control would seriously narrow scientific vision and "cut the canvas" of the scientific worker. I feel, however, that while this is true, such steps are necessary in the initial stages and the research perspective can widen as scientific and technological independence is attained. Some of the scientific priority areas are discussed in a later section of this article.

It is also beyond the scope of this article to discuss the history of materialism. Broadly speaking there are two main streams of materialistic philosophy that the Muslim countries have to counter today. The first consists of non-Marxist Western atheists who scorn religion and, dazzled by the superficial glamour of the West, would like to see a blind adoption of Western culture and social attitudes which today have lost all sense of morality. Indeed words such as "kindness", "conscience" and "morality" have little meaning for them as they indulge in the excesses that the Western way of life allows them. All sense of decency and morality is sacrificed at the altar of "progress" and "modernization". Many of these "progressives" occupy seats of power in various governments and hoodwink the masses by wearing one of the "Islamic garbs". The most potent weapon in the demolition of Islamic values has indeed been the exploitation of the masses in the name of "Islam", "Islamic socialism" or "Islamic democracy", in the various Muslim countries of the world. While dressing, living and drinking like the Westerner, these leaders dupe the people by means of Islamic slogans in order to achieve greedy and selfish ends of lust and power. Under their regimes the moral decay which started under colonial domination has accelerated tremendously as one individual tries to outdistance another in the race for worldly gain, while branches of the media are fully controlled and exploited to eradicate the practice of Islam from the Muslim world. Apparently harmless Western films displayed on television gradually but surely erode all Islamic values and "Westernize" the masses, and render acceptable codes of conduct which would otherwise have been unacceptable. In fact television and the cinema, which could have proved to be extremely powerful tools in education, in the inculcation of the Islamic spirit of Jihad and the unification of the Muslim world, are being callously exploited to accelerate moral decay. What is the result? A group of isolated Muslim nations, spreading from North Africa in the West through the Middle East to the Far East, all wallowing in their petty narrow-minded power politics, are searching for so-called "Progress".

In those Muslim countries which were colonized by the Western powers, the Christian missionary schools have contributed a great deal to the perpetration of anti-Islamic bias. Being unable to convert children to Christianity, they did the next best thing in seducing their minds away from Islam. In such countries missionary schools provided a high standard of "secular" education and therefore tended to attract children of the most influential people in the community. Educated in a Western materialistic way of life, these children then become the leaders of the community and the process is therefore self-perpetuating. This system was initiated in India, for instance, through a British education policy which resulted in the closing of "madrassas" and the establishment of the British school system. Thus the very roots of Muslim culture were cut down. In Turkey the forces of Westernization almost completely destroyed its Muslim character and today Turkey prides itself on being "European". Foreign languages such as English and French became the medium of instruction at school, college and university level, thus isolating students from the vast body of literature in their own languages. The intelligentsia which emerged from these universities were secularly trained in their fields and the burning passion for Islamic tabligh and Jihad that once enlightened the entire East was sadly missing. Trained by teachers with Western outlooks, these students were ill-prepared in the moral and intellectual values necessary in a Mussulman. As a result, the loss of the Islamic heritage has led to a situation in which the practice of Islam appears as a meaningless dogmatic ritual to the leaders of Muslim society, which itself has become rootless and ineffective. The second of the principal materialistic forces at work in Muslim countries is that with a Marxist outlook. Marxists, too, had to wear an Islamic garb in many Muslim countries in order to propagate their evil beliefs. Inspired by the communist blocs, these materialists have gone about destroying all religious values wherever they have gained power or have been in a position to do so. The process has been helped by holding out promises of equality for all, especially in the less developed countries of the world where the frustrated poor masses are prone to be easy prey to such cleverly contrived propaganda.

The result of this movement away from Islam in the Muslim world has been accompanied by a loosening of moral fibre, a consequent increase in the rate of crime and a lack of willingness to accept social responsibilities. Take away education and the moral codes of religion from the masses and what is left? A collection of intelligent animals who live from day to day without any concept of right or wrong, satisfying their passions and inner urges at the expense of physical, mental and moral well-being. The luckier ones amongst these in the Western world adopt some sense of social responsibility from the education they receive but are still susceptible to the variety of psychological diseases that accompany a lack of faith. As a result the suicide rates in the Western world are the highest in the world — a civilized world or a lonely jungle?

While one may accept the validity of these criticisms, the question still arises: what has all this to do with teaching science in the Muslim world? Are these not social problems which must be solved separately? Should not science remain divorced from such political considerations and concentrate purely on the search for material truth? Does science not offer an ethic of its own which could possibly become an ethic for all humanity? Many students of science believe that this is so and choose to live their lives oblivious to other considerations. The above arguments may on superficial examination appear attractive to some but they cannot withstand a more searching probe. Can science stay divorced from resulting social implications? The "pursuit" of truth unrelated to social obligations can lead to a devastating end for all humanity. There is a growing number of areas in science where the truth of this is clearly visible. The field of genetical engineering is one. The ability to manipulate the human species so as to be able to breed geniuses, labourers or soldiers has horrific possibilities. The power of the atom for destructive purposes has already been demonstrated in Hiroshima and Nagasaki. Research in the fields of nerve gas and lethal germs have already resulted in the development of horrifying destroyers of humanity, many of which were freely employed in Vietnam. A field of emotional chemistry that is evolving may eventually make it possible to control the human "will" through chemical agents. In all these developments scientists play a crucial role. However, they are not alone to blame. In almost all cases, efforts in such "military" areas are government-directed and the scientists are no more than pawns on the international chess board. The point that emerges is that the "pursuit of knowledge", however glorified the words may sound, can have consequences of a seriously detrimental nature unless the precise boundaries of "knowledge" are more clearly defined. The fathomless depths of the beauty and wonder of nature offer vistas for human endeavour which can lead man in any number of directions of scientific research. It is up to the leaders of the scientific community first to examine in depth the problems faced by humanity, such as disease, famine, and poor living standards and then to direct the efforts of scientists towards solving these pressing problems.

Again one encounters the valid argument that if the leading Western and communist countries are concentrating on the development of extremely sophisticated war machinery, should the Muslim countries stand by and for the sake of "ethical considerations" avoid working in such scientific areas. Clearly this is not possible unless there is international disarmament, as our very survival depends on our ability to defend ourselves. So far this defence has rested on the "charity of armaments" that the Western countries have doled out. The lack of education, technological know-how and experience in the manufacture of sophisticated weapons has meant that a beggar's bowl has constantly been in the hands of most Muslim states, and those who follow American or communist political lines have been rewarded by the presentation of aircraft, tanks and missiles. This has however all been done with extreme caution by the world powers, who make sure that predestined strategic balances or imbalances are maintained. No serious attempt has been made in the Muslim world to achieve technological independence, and no doubt the major world powers will do their best to prevent this from happening. American political pressure on France to prevent the setting up of a nuclear fuel re-processing plant in Pakistan is a striking example of such pressures. Because they lack faith themselves and are overawed by Western technology Muslim nations have made little or no attempt to establish an independent, sophisticated arms industry; consequently many of them are having to pay 50 to 100 times the cost of manufacture of a guided missile system, for example, in order to acquire it. The fact that it requires very little effort to make a duplicate of even the most complicated electronically guided system, at only 1%–5% of the cost of the original, does not appear to have occurred to these states. Oil wealth is today therefore benefiting the West far more than it is helping the Arabs who conveniently exchange this extremely valuable source of petrochemicals for a few tanks and aeroplanes. The rapid exhaustion of this raw material and the accelerated pace of research in nuclear fusion technology will undoubtedly lead to the replacement of oil by thermonuclear energy derived from sea water as the primary source of energy by the turn of the century and it is up to Muslims to see that they are not left behind in this vital area of research. Research in the Muslim world towards the establishment of commercial nuclear fusion reactors must begin immediately. These are some of the pitfalls which we must avoid in designing our own educational system and I am listing here some measures that need to be imposed in Muslim states to ensure that they achieve technological equivalence with the West without compromising their Islamic ideology etc.

Scientific education in Muslim countries should be based on principles and guidelines laid down in the Quran and Hadith, and the following steps should be taken to achieve this objective:

1. The end-products of the system of education should be men of outstanding calibre excelling in their own fields of learning but at the same time conforming to the Islamic ideology in their thoughts and deeds. Courses should include a comprehensive programme of religious education at school, colleges and university clearly defining the limits of scientific knowledge and clarifying how modern technological advances fit in with Islamic values so that the evils of materialism can be avoided.

2. Ways and means should be devised for preventing the insidious aspects of Western civilization from penetrating our society. The steps to be taken should include:

   (a) The banning of Western films except those of educational value;
   (b) The strict punishment of drug abuses;
   (c) An organized programme fully using the media, i.e. television, radio and the newspapers, to educate the masses in religious and scientific knowledge;
   (d) The strict punishment of the sale of pornographic literature;
   (e) The banning of the import of alcoholic drinks;
   (f) The closing of all night clubs.

3. There should be comprehensive religious education at school, college and university by teachers who are properly qualified for the job. A degree in religious knowledge together with subjects such as the natural sciences, history and philosophy should be compulsory for all religious instructors so that they can correlate and explain the relevance of Islam in this fast-changing modern world to today's genuinely curious students who wish to base their beliefs on logic rather than on blind faith.

4. Arabic should be taught as a compulsory subject at school, college and university in all Muslim countries.

5. Radio, television and the newspapers should be used primarily for educational purposes to inculcate the spirit of Islam in the masses, educational programmes to comprise 80% of total viewing time. It is suggested that film recordings of lectures by eminent scientists in various

areas should be prepared and transmitted on video-cassette recorders installed in every college or university department. This would make it unnecessary for a lecturer to deliver the same lecture every year and would provide him with enough time to keep abreast of the latest developments in his subject, as well as to give individual personal attention to the problem which his students may have. The recordings could be cumulative as the subject evolved each year.

6. A programme should be organized among Muslim nations to encourage the writing of scientific books and papers.

7. A multi-million-rial research-sponsoring organization should be established which would carefully choose its priority areas in order to achieve the following:
   (a) Technological independence from the advanced countries;
   (b) The setting up of giant industrial projects;
   (c) The manufacture of ships, submarines, aircrafts, rockets, missiles, tanks, explosives and other war machinery for which the Muslims have so far had to rely on the West; and
   (d) Other priority areas to be decided by discussion.

8. The framework of the proposed United States of Islam should be set up. Steps should be taken to convey to the masses in various Muslim countries that their nationalistic feelings are as artificial as the borders between their countries, and to emphasize the need for the unification of all Islamic countries in one nation governed by a Council of men of outstanding learning.

9. Technological universities should be established in each Muslim country, sponsored by a central fund and equipped with the most sophisticated tools of research; the maximum research efforts should be made in certain high priority areas specified by the organizers of the Research Fund.

10. Immediate steps should be taken to create a close-knit coordinated programme of scientific and technological development in Muslim countries so that they can achieve complete technological independence from advanced Western nations within a specific time-limit (provisionally ten years) and can become leaders in the areas specified below in the subsequent five years:
   (a) Electronics: A coordinated system of manufacture of electronic

components ranging from simple instruments to the most sophisticated systems such as laser-guided missiles and solid state circuitry;

(b) Chemical Industry: The manufacture of industrial items allied to the petrochemical industry e.g. synthetic fibres, explosives and pharmaceuticals;

(c) Metallurgy: Research in the fields of metallurgy for the purpose of obtaining metals for the aircraft industry, tanks and automobiles. Setting up steel mills to cope with the requirements of the entire Muslim world;

(d) Aircraft Industry: Setting up a multinational aircraft industry designed to produce the most sophisticated aircraft, missiles with various guided systems;

(e) Transport Industry: The manufacture of automobiles, trains and ships catering for the entire Muslim world;

(f) Other miscellaneous industries, e.g. cement and fertilisers necessary for independence from Western materials.

11. The employment of some of the leading experts in their fields (if necessary from U.S.A., Canada or Europe) by offering them lucrative salaries in order to launch research and industrial programmes in the fields of electronics, metallurgy, chemicals and the aircraft industry. These experts should be employed only for a period of 5–7 years and replaced by Muslim scientists within that period.

12. Programmes of exchange visits between outstanding educationists of Muslim countries with the aim of creating a greater harmony and understanding between them.

13. The establishment of the world's finest libraries in each Muslim country, fully equipped to provide any information in the form of reprints of articles within seven days of the receipt of a request (again to be financed from a central fund).

14. The construction of large nuclear reactors without Western know-how leading to the independent production of atomic weapons.

15. The establishment of research centres in areas of nuclear fission, and more importantly, in nuclear fusion technology, leading to the independent establishment of nuclear fusion reactors as sources of energy.

16. The inclusion in educational courses of works of great Muslim scholars in various subjects so as to highlight the glorious heritage that we have. The same policy should be implemented in literature where

the works of Muslim scholars, or translations of them, should be read as standard course books, to replace the present emphasis on Western writers such as Shakespeare and Dickens.

17. The fostering of a closer awareness of the works of Muslim scientists such as Ibne Sina, Khwarzeme, Ibne Rushd, Kindi, Abu Mohammad Khajundi, Jabir Bin Hayyan and Mosa Bin Shakir.

18. The inclusion in college/university courses of a study of the contributions of Muslim economists such as Ibne Khaldun and Shah Waliyyullah. Similarly the works of Muslim scholars in the fields of political science, sociology, psychology and jurisprudence should be prescribed.

19. Every university to have full faculties of Islamic Studies and every student taking a university degree to be made to achieve a certain qualifying level in Islamyat. This should include topics like "Ilme Aqqayad", "Ilme Hadith", "Ilme Usule Hadith", "Ilme Tafsir", "Ilme Usule Tafsir", "Feqah" and "Ilme Usule Fiqah". The teachers of Islamiyat on the other hand, besides holding a first class M.A. in Islamiyat, must have passed general science courses of a high standard.

20. A system to be established for weeding out non-productive teachers. This should be done by abolishing permanent posts, and introducing a five-year tenure system. At the end of the fourth year each teacher should be required to apply for a renewal of tenure and to give evidence of his productivity in terms of research publications or books. These should be evaluated for extent and quality by eminent scholars from another Muslim country who would decide whether the tenure of the applicant should be renewed or not.

21. The environment of Muslim educational institutions to project their Islamic character. A simple Islamic dress should be made compulsory and Western dresses and styles should be forbidden.

22. The introduction of general science courses for all students in faculties other than the science faculty.

23. Friday to be introduced as a holiday, instead of Sunday, in all educational institutions.

All these developments are possible only, of course, if Islam-loving scholars are prepared to be responsible for bringing about the changes. Only persons eminent in their subject and known for their Islamic faith should be put in charge of the committees which are formed to do so.

STAFFORD LIBRARY
COLUMBIA COLLEGE
1001 ROGERS STREET
COLUMBIA, MO 65216